11th International Workshop on Health Text Mining and Information Analysis (LOUHI 2020)

Online
20 November 2020

ISBN: 978-1-7138-1997-4

EMNLP 2020

The 11th International Workshop on Health Text Mining and Information Analysis
LOUHI 2020

Proceedings of the Workshop

November 20, 2020

Introduction

The International Workshop on Health Text Mining and Information Analysis (LOUHI) provides an interdisciplinary forum for researchers interested in automated processing of health documents. Health documents encompass electronic health records, clinical guidelines, spontaneous reports for pharmacovigilance, biomedical literature, health forums/blogs or any other type of health-related documents. The LOUHI workshop series fosters interactions between the Computational Linguistics, Medical Informatics and Artificial Intelligence communities. The 10 previous editions of the workshop were co-located with SMBM 2008 in Turku, Finland, with NAACL 2010 in Los Angeles, California, with Artificial Intelligence in Medicine (AIME 2011) in Bled, Slovenia, during NICTA Techfest 2013 in Sydney, Australia, co-located with EACL 2014 in Gothenburg, Sweden, with EMNLP 2015 in Lisbon, Portugal, with EMNLP 2016 in Austin, Texas; in 2017 was held in Sydney, Australia; in 2018 was co-located with EMNLP 2018 in Brussels, Belgium; and in 2019 was co-located with EMNLP 2019 in Hong Kong. This year the workshop is co-located with EMNLP 2020 and takes place online due to the COVID-19 pandemics.

The aim of the LOUHI 2020 workshop is to bring together research work on topics related to health documents, particularly emphasizing multidisciplinary aspects of health documentation and the interplay between nursing and medical sciences, information systems, computational linguistics and computer science. The topics include, but are not limited to, the following Natural Language Processing techniques and related areas:

- Techniques supporting information extraction, e.g. named entity recognition, negation and uncertainty detection

- Classification and text mining applications (e.g. diagnostic classifications such as ICD-10 and nursing intensity scores) and problems (e.g. handling of unbalanced data sets)

- Text representation, including dealing with data sparsity and dimensionality issues

- Domain adaptation, e.g. adaptation of standard NLP tools (incl. tokenizers, PoS-taggers, etc) to the medical domain

- Information fusion, i.e. integrating data from various sources, e.g. structured and narrative documentation

- Unsupervised methods, including distributional semantics

- Evaluation, gold/reference standard construction and annotation

- Syntactic, semantic and pragmatic analysis of health documents

- Anonymization/de-identification of health records and ethics

- Supporting the development of medical terminologies and ontologies

- Individualization of content, consumer health vocabularies, summarization and simplification of text

- NLP for supporting documentation and decision making practices

- Predictive modeling of adverse events, e.g. adverse drug events and hospital acquired infections

- Terminology and information model standards (SNOMED CT, FHIR) for health text mining

- Bridging gaps between formal ontology and biomedical NLP

The call for papers encouraged authors to submit papers describing substantial and completed work but also focus on a contribution, a negative result, a software package or work in progress. We also encouraged to report work on low-resourced languages, addressing the challenges of data sparsity and language characteristic diversity.

This year we received a high number of submissions (43), therefore the selection process was very competitive. Due to time and space limitations, we could only choose a small number of the submitted papers to appear in the program.

Each submission went through a double-blind review process which involved three program committee members. Based on comments and rankings supplied by the reviewers, we accepted 16 papers. Although the selection was entirely based on the scores provided by the reviewers, we regretfully had to set a relatively high threshold for acceptance. The overall acceptance rate is 37%.

Our special thanks go to Guergana Savova for accepting to give an invited talk.

Finally, we would like to thank the members of the program committee for providing balanced reviews in a very short period of time, and the authors for their submissions and the quality of their work.

Organizers:

Eben Holderness, Brandeis University, USA
Antonio Jimeno Yepes, IBM Research, Australia
Alberto Lavelli, FBK, Trento, Italy
Anne-Lyse Minard, University of Orleans, France
James Pustejovsky, Brandeis University, USA
Fabio Rinaldi, Dalle Molle Institute for Artificial Intelligence Research - IDSIA, University of
Zurich, Switzerland & FBK, Trento, Italy

Program Committee:

Mohammad Akbari, National University of Singapore, Singapore
Rafael Berlanga Llavori, Universitat Jaume I, Spain
Georgeta Bordea, Université de Bordeaux, France
Leonardo Campillos Llanos, LIMSI, CNRS, France
Francisco Couto, University of Lisbon, Portugal
Hercules Dalianis, Stockholm University, Sweden
Kerstin Denecke, Bern University of Applied Sciences, Switzerland
Martin Duneld (xmartin@dsv.su.se), Stockholm University, Sweden
Natalia Grabar, CNRS UMR 8163, STL Université de Lille3, France
Cyril Grouin, LIMSI, CNRS, Université Paris-Saclay, Orsay, France
Thierry Hamon, LIMSI, CNRS, Université Paris-Saclay, Orsay, France & Université Paris 13, Vil-
letaneuse, France
Aron Henriksson, Stockholm University, Sweden
Eben Holderness, Brandeis University, USA
Rezarta Islamaj-Dogan, NIH/NLM/NCBI, USA
Hyeju Jang, University of British Columbia, Canada
Antonio Jimeno Yepes, IBM Research, Australia
Yoshinobu Kano, Shizuoka University, Japan
Jin-Dong Kim, Research Organization of Information and Systems, Japan
Dimitrios Kokkinakis, University of Gothenburg, Sweden
Martin Krallinger, Spanish National Cancer Research Centre (CNIO), Spain
Alberto Lavelli, FBK, Trento, Italy
Analia Lourenco, Universidade de Vigo, Spain
David Martinez, University of Melbourne and MedWhat.com, Australia
Sérgio Matos, University of Aveiro, Portugal
Timothy Miller, Harvard Medical School, USA
Anne-Lyse Minard, University of Orleans, France
Hans Moen, University of Turku, Finland
Roser Morante, VU Amsterdam, Netherlands
Danielle L Mowery, University of Utah, USA
Henning Müller, University of Applied Sciences Western Switzerland, Switzerland
Aakanksha Naik, CMU, USA
Aurélie Névéol, LIMSI, CNRS, Université Paris-Saclay, Orsay, France
Mariana Lara Neves, German Federal Institute for Risk Assessment, Germany
Jong C. Park, KAIST Computer Science, Korea
Laura Plaza, Universidad Complutense de Madrid, Spain

James Pustejovsky, Brandeis University, USA
Fabio Rinaldi, Dalle Molle Institute for Artificial Intelligence Research - IDSIA, University of Zurich, Switzerland & FBK, Trento, Italy
Thomas Brox Røst, Norwegian University of Science and Technology, Norway
Tapio Salakoski, University of Turku, Finland
Stefan Schulz, Graz General Hospital and University Clinics, Austria
Maria Skeppstedt, Linneus University, Sweden, and Potsdam University, Germany
Amber Stubbs, Simmons College, USA
Hanna Suominen, Australian National University, Australia
Suzanne Tamang, Stanford University School of Medicine, USA
Sumithra Velupillai, KTH, Royal Institute of Technology, Sweden, and King's College London, UK
Özlem Uzuner, MIT, USA
Yanshan Wang, Mayo Clinic, USA
Pierre Zweigenbaum, LIMSI, CNRS, Université Paris-Saclay, Orsay, France

Additional Reviewers:

Marcia Barros, University of Lisbon, Portugal
Jari Björne, University of Turku, Finland
Juho Heimonen, University of Turku, Finland
Huije Lee, KAIST Computer Science, Korea
Kahyun Lee, MIT, USA
Pedro Ruas, University of Lisbon, Portugal
Hoyun Song, KAIST Computer Science, Korea
Diana Sousa, University of Lisbon, Portugal
Wonsuk Yang, KAIST Computer Science, Korea

Invited Speaker:

Guergana Savova, Boston Children's Hospital and Harvard Medical School, USA

Table of Contents

Conference Program

November 20, 2020

09:00–10:45   **Session 1**

9:00   *Introduction*

09:15   *The Impact of De-identification on Downstream Named Entity Recognition in Clinical Text*
Hanna Berg, Aron Henriksson and Hercules Dalianis

09:30   *Simple Hierarchical Multi-Task Neural End-To-End Entity Linking for Biomedical Text*
Maciej Wiatrak and Juha Iso-Sipila

09:40   *Medical Concept Normalization in User-Generated Texts by Learning Target Concept Embeddings*
Katikapalli Subramanyam Kalyan and Sivanesan Sangeetha

09:50   *Not a cute stroke: Analysis of Rule- and Neural Network-based Information Extraction Systems for Brain Radiology Reports*
Andreas Grivas, Beatrice Alex, Claire Grover, Richard Tobin and William Whiteley

10:05   *GGPONC: A Corpus of German Medical Text with Rich Metadata Based on Clinical Practice Guidelines*
Florian Borchert, Christina Lohr, Luise Modersohn, Thomas Langer, Markus Follmann, Jan Philipp Sachs, Udo Hahn and Matthieu-P. Schapranow

10:20   *Session 1 QA*

10:45–11:00 Break

11:00–12:20 Session 2

11:00   *Normalization of Long-tail Adverse Drug Reactions in Social Media*
Emmanouil Manousogiannis, Sepideh Mesbah, Alessandro Bozzon, Robert-Jan Sips, Zoltan Szlanik and Selene Baez

11:15   *Evaluation of Machine Translation Methods applied to Medical Terminologies*
Konstantinos Skianis, Yann Briand and Florent Desgrippes

11:30   *Information retrieval for animal disease surveillance: a pattern-based approach.*
Sarah Valentin, Mathieu Roche and Renaud Lancelot

11:45   *Multitask Learning of Negation and Speculation using Transformers*
Aditya Khandelwal and Benita Kathleen Britto

12:00   *Session 2 QA*

12:20–14:00 Break

14:00–14:45 Invited Talk

14:00   *TBA*
Guergana Savova

14:30   *Invited Talk QA*

14:45–15:00   **Break**

15:00–16:15   **Session 3**

15:00   *Biomedical Event Extraction as Multi-turn Question Answering*
Xing David Wang, Leon Weber and Ulf Leser

15:15   *An efficient representation of chronological events in medical texts*
Andrey Kormilitzin, Nemanja Vaci, Qiang Liu, Hao Ni, Goran Nenadic and Alejo
Nevado-Holgado

15:25   *Defining and Learning Refined Temporal Relations in the Clinical Narrative*
Kristin Wright-Bettner, Chen Lin, Timothy Miller, Steven Bethard, Dmitriy Dligach, Martha Palmer, James H. Martin and Guergana Savova

15:40   *Context-Aware Automatic Text Simplification of Health Materials in Low-Resource
Domains*
Tarek Sakakini, Jong Yoon Lee, Aditya Duri, Renato F.L. Azevedo, Victor
Sadauskas, Kuangxiao Gu, Suma Bhat, Dan Morrow, James Graumlich, Saqib
Walayat, Mark Hasegawa-Johnson, Thomas Huang, Ann Willemsen-Dunlap and
Donald Halpin

15:55   ***Session 3 QA***

16:15–16:30   **Break**

16:30–17:30   **Session 4**

16:30   *Identifying Personal Experience Tweets of Medication Effects Using Pre-trained
RoBERTa Language Model and Its Updating*
Minghao Zhu, Youzhe Song, Ge Jin and Keyuan Jiang

16:45   *Detecting Foodborne Illness Complaints in Multiple Languages Using English Annotations Only*
Ziyi Liu, Giannis Karamanolakis, Daniel Hsu and Luis Gravano

17:00   *Detection of Mental Health from Reddit via Deep Contextualized Representations*
Zhengping Jiang, Sarah Ita Levitan, Jonathan Zomick and Julia Hirschberg

17:15   ***Session 4 QA***

The Impact of De-identification on Downstream Named Entity Recognition in Clinical Text

Hanna Berg, Aron Henriksson, Hercules Dalianis
Department of Computer and Systems Sciences
Stockholm University, Sweden
`{hanna.berg,aronhen,hercules}@dsv.su.se`

Abstract

The impact of de-identification on data quality and, in particular, utility for developing models for downstream tasks has been more thoroughly studied for structured data than for unstructured text. While previous studies indicate that text de-identification has a limited impact on models for downstream tasks, it remains unclear what the impact is with various levels and forms of de-identification, in particular concerning the trade-off between precision and recall. In this paper, the impact of de-identification is studied on downstream named entity recognition in Swedish clinical text. The results indicate that de-identification models with moderate to high precision lead to similar downstream performance, while low precision has a substantial negative impact. Furthermore, different strategies for concealing sensitive information affect performance to different degrees, ranging from pseudonymisation having a low impact to the removal of entire sentences with sensitive information having a high impact. This study indicates that it is possible to increase the recall of models for identifying sensitive information without negatively affecting the use of de-identified text data for training models for clinical named entity recognition; however, there is ultimately a trade-off between the level of de-identification and the subsequent utility of the data.

1 Introduction

There is a growing demand for access to large amounts of healthcare data in order to facilitate research and development of tools for healthcare management and clinical decision support, not least as a result of the increasing application of AI and machine learning in healthcare. However, to enable large-scale secondary use of sensitive healthcare data, there is a need for automatic privacy-protecting methods for clinical text; manual de-identification to ensure that the data does not contain personal information is often prohibitively expensive.

Privacy-protecting methods for de-identification of data generally address three privacy risks: (i) the risk that someone's records can be uniquely identified in a dataset, (ii) preventing linkage from one dataset to another, and (iii) the risk of inferring sensitive information about an individual from the dataset (EU, 2014). De-identification of unstructured clinical text commonly focuses on identifying potentially identifiable information within predefined classes in an approach based on named entity recognition (NER), where protected health information (PHI) is first identified and subsequently obscured in some fashion (Meystre et al., 2010). This technique addresses the first of the mentioned privacy risks, and is the focus of this study.

De-identification techniques for structured data generally address all three mentioned privacy risks, and their impact on data quality and data utility are more thoroughly studied than the impact of de-identification on unstructured text (Iwuchukwu et al., 2007). Research on structured data shows that de-identification techniques applied to structured data may lead to reduced data quality (Xia et al., 2015). For both structured and unstructured data, information necessary for answering a research question may be identifying, and therefore required to be removed or altered to ensure the privacy of data subjects. Furthermore, for automatic de-identification, relevant information may be changed or removed unintentionally due to non-sensitive tokens being misclassified as sensitive. Studies have indicated that de-identification may erroneously alter or hide data significant for other tasks; however, no significant impact has so far been observed when comparing models for downstream tasks trained with original versus de-identified text data (Deleger et al., 2013; Obeid et al., 2019; Meystre et al., 2014). It is, however,

1

Proceedings of the 11th International Workshop on Health Text Mining and Information Analysis, pages 1–11
November 20, 2020. ©2020 Association for Computational Linguistics
https://doi.org/10.18653/v1/P17

still not clear how the impact is related to various forms and levels of de-identification, in particular the trade-off between the precision and recall of the de-identification system.

Common measures for evaluating de-identification models for unstructured data are precision, recall and F_1-score. High recall is generally preferred over high precision, as the privacy of the data subjects is prioritised over potential loss of document interpretability (Ferrández et al., 2012). There is, at the same time, a concern that poor precision would have a negative impact on the quality of the data. In practice, precision and recall are typically evaluated as equally important when using F_1-score as the primary evaluation metric. However, if precision of the de-identification model is of less importance and it turns out that this does not have a clear impact on using the data for some downstream task, it would entail that de-identification systems can be adapted for high coverage – but also that recall should carry more weight than precision when evaluating de-identification systems.

On the other hand, if de-identification impacts the text data quality negatively, the concern is that it would lower the possibility to use the data for building models for various downstream tasks, i.e. data utility would decrease once it has been de-identified. Examples of downstream tasks are detection of healthcare-associated infections, adverse drug events and early cancer symptoms.

This study intends to answer the question of how the precision of a de-identification system affects data quality and, in particular, data utility by investigating how different levels of de-identification affects the performance of downstream clinical NER. In particular, the study aims to investigate the impact of the trade-off between precision and recall, as well as different methods for concealing PHIs.

2 Related Research

To our knowledge, no studies have specifically focused on the trade-off between precision and recall and its impact on downstream tasks. There are, however, studies that have investigated the impact of de-identification.

A study by Meystre et al. (2014) showed that de-identification reduces the information content, and leads to the possible introduction of misleading information if tokens are replaced with pseudonyms. In the study, only 0.81% of clinical named entity annotations were erroneously detected as PHI, but between 10 and 49% of all eponyms[1] were misclassified as PHI. Fewer SNOMED-CT concepts were also found, but this was largely explained by incorrectly labelled SNOMED-CT concepts in the original dataset. Studies on downstream tasks have, however, not shown that de-identification has a negative impact. No significant differences could be seen between using original text or de-identified text as training data for clinical text classification (Obeid et al., 2019). In contrast, one study observed potentially significant benefits of training on de-identified data for medication name extraction (Deleger et al., 2013). The reduction of dimension in de-identified text has been hypothesised to potentially improve machine learning performance (Obeid et al., 2019).

While previous studies have compared different systems with various levels of precision (Obeid et al., 2019; Meystre et al., 2014), no study has explicitly compared the impact the precision of a de-identification system has on downstream tasks. In Deleger et al. (2013), a recall bias was introduced in two versions and those models were compared to the original version. The systems with a slightly lower precision performed similarly to the original one in terms of performance on a medication name extraction task.

3 Methods and Materials

In this paper, the impact of various forms of de-identification – where a trade-off is made between precision and recall of the model used for identifying PHI, as well as using different levels of de-identification – on downstream NER tasks, in this case identifying various clinical entity types, is studied.

This paper uses four corpora: one for development of de-identification models and three for evaluating the impact of de-identification on downstream clinical NER. The method is presented first, followed by a description of the corpora. The experiments include five models for identifying PHI, each one trading off precision for increased recall to different extents, as well a four concealment strategies that hide the identified PHI to different degrees. These PHI models and concealment strategies are used to de-identify three clinical corpora[2],

---

[1]Eponyms are terms named after researchers, such as *Crohn's disease*, *Cushing's syndrome* or *Waldenström macroglobulinemia*.

[2]This research has been approved by the Swedish Ethical

which, in turn, are used for training downstream clinical NER models. The impact of using different PHI models and concealment strategies is then analysed from a number of different perspectives.

3.1 Methods

The impact of PHI models and concealment of PHI is evaluated based on their performance as training data for downstream clinical NER tasks. The overlap and co-occurrence between manually annotated clinical entities and predicted PHI are also analysed to find out if certain PHI classes have a bigger impact on the downstream clinical NER tasks. Furthermore, the impact of de-identification on eponyms are analysed on a new corpus specifically made for this purpose.

3.1.1 De-identification Process

The de-identification process consists of two main steps: (i) identification of PHI and (ii) concealment of PHI.

3.1.2 Optimising the F-score

Optimisation for different F-scores was carried out to obtain models with higher recall at the expense of precision, with the aim of investigating the effect this has on downstream clinical NER. Deleger et al. (2013) introduced a recall bias by changing the predicted label of non-PHI tokens with system-generated probability less or equal to a threshold of 0.95 to the PHI label with the second highest probability. In this study, this was done in a similar fashion by changing predicted non-PHI labels to their most likely alternative PHI label if the the marginal score was lower than the set threshold. The marginal probability specifies the model's confidence in predicting each label of an input sequence, without regard to the outcome of other variables (Sutton and McCallum, 2012).

The thresholds were decided through grid search, optimising toward five different F-scores: F_1, F_4, F_{10}, F_{20} and F_{40}. F-score is the weighted mean of precision and recall, see Eq. 1. For F_1, equal weight is given to precision and recall. For F_2, recall is given twice the weight of precision. Hence, β will obtain the following values: 1, 4, 10, 20 and 50.

$$F\text{-}score : F_\beta = (1 + \beta^2) * \frac{P * R}{\beta^2 * P + R} \quad (1)$$

Review Authority under permission no. 2019-05679.

3.1.3 Concealment Strategies

The next step is to conceal the PHI. The four concealment strategies used in this study are: *Pseudo*, *Class*, *Mask* and *Remove*.

- *Original* – No de-identification method. Example: "Eva slept."
- *Pseudo* – To replace the identified PHI with a surrogate. Example: "Mary slept."
- *Class* – To replace the identified PHI with the PHI class. Example: "<First_Name> slept."
- *Mask* – To replace the identified PHI with *XXXX*. Example: "XXXX slept."
- *Remove* – To completely remove the identified PHI and the sentence it is contained. Example: " "

The different methods hide information to varying degrees. *Pseudo* replaces some information, but, for example, may retain information about time between different events by shifting dates consistently. For *Class*, there is still information about the type of PHI found, which is missing for *Mask*. *Remove* removes not only the PHI itself but also the context surrounding it, i.e. all sentences where a PHI is found are removed.

The pseudonymisation algorithm was similar to the one described in (Dalianis, 2019; Berg et al., 2019). The changes are based on the error analysis in (Berg et al., 2019). The changes are:

- Uncommon names are replaced with uncommon names.
- The number of tokens of each PHI instance is kept.
- The format of dates are kept.
- Tokens for *Health Care Units* are replaced with the same strategy as the one used for replacing. *Locations* in Dalianis (2019).
- Not all *Health Care Units* tokens are replaced, but at least one token within every entity. Locations and named entities are replaced, while for example tokens like "hospital" or "clinic" are not replaced,

For each clinical NER corpus, 25 dataset variants were produced. The datasets had six levels of de-identification based on f-scores, ranging from not de-identified to increasingly larger percentages of both true and false positives. The identified PHI were then managed using four different concealment strategies: pseudonymising the PHI (*Pseudo*), replacing the PHI with the classified PHI type (*Class*), masking the PHI (*Mask*) or removing whole sentences containing PHI (*Remove*). This resulted in 24 de-identified datasets and one original dataset for training.

3.2 Materials

All corpora are contained in the research infrastructure Health Bank – the Swedish Health Record Research Bank[3]. Health Bank contains electronic patient records from over two million patients from Karolinska University Hospital from the years 2006–2014.

PHI Corpus

The de-identification models was trained, as well as evaluated, on the Stockholm EPR PHI Corpus (Velupillai et al., 2009; Dalianis and Velupillai, 2010). The Stockholm EPR PHI Corpus consists of 98 patient records in Swedish from five clinical units at Karolinska University Hospital: *neurology, orthopaedia, infection, dental surgery* and *nutrition*. The corpus contains nearly 200,000 tokens in total. The annotated classes are: *Age, Full Date, Date Part, First Name, Last Name, Health Care Unit, Location* and *Phone Number* distributed over 4,826 annotated entities. The dataset includes free text and information about which section the text is from. The distribution is imbalanced with roughly 3% of all tokens being part of an annotated entity.

Four out of five clinical units were used for training and the tuning of the hyper parameters. The data from the *neurology clinical unit* was used for the final evaluation.

Clinical Entity Recognition Corpora

The following three corpora were used for building and evaluating clinical NER models, with and without de-identification.

Stockholm EPR Clinical Entity Corpus is a corpus in Swedish for clinical NER with clinical notes from an internal medicine emergency unit at Karolinska University Hospital (Skeppstedt et al., 2014). The corpus was annotated by three annotators; the annotation process and the resulting corpus are described in (Skeppstedt et al., 2014; Kvist et al., 2011). The dataset is annotated with the labels: *Explicit Disorder, Implicit Disorder, Finding, Drug* and *Body structure* distributed over totally 7,946 annotated entities.

Stockholm EPR (Adverse Drug Event) ADE Corpus is a corpus in Swedish for adverse drug events, annotated with clinical named entities (Henriksson et al., 2015). The clinical notes are extracted from the larger Health Bank by extracting data with ICD-10 codes marking an adverse drug event. The dataset uses the labels *Disorder, Finding, Drug, ADE Cue* and *Body structure* distributed over totally 3,789 annotated entities.

Stockholm EPR Cervical Cancer Corpus is a corpus in Swedish with clinical records for patients with a cervical cancer diagnosis. The texts are annotated with the labels *Finding, Disorder* and *Body part*, distributed over totally 7,663 annotated entities. The annotation process is described in (Weegar et al., 2015).

Eponym Corpus

An additional corpus was constructed in order to enable a more thorough analysis of the impact on eponyms. The dataset is a subset of a larger corpus extracted from Health Bank, with over 213 million tokens. For each of the 12 most common eponyms, one hundred sentences in which the eponym appeared were extracted.

3.3 Experimental setup

In the analyses of the clinical NER corpora, *First Name* and *Last Name* are merged to *Person* and *Date Part* and *Full Date* to *Date*. *Explicit disorder, Implicit Disorder* and *Disorder* are similarly all merged to one category. In the analyses, we present results averaged over the three corpora.

The PHI models are evaluated on the subset of Stockholm EPR PHI Corpus that they are not trained on. The main evaluation is a binary evaluation to investigate how many non-PHI that are classified as PHI. The evaluation is also token-based, meaning that they are evaluated on a per-token basis. The system's ability to locate where an PHI begins and another ends is not of great importance for the replacement step, but instead the focus is on whether tokens are replaced or not.

For each clinical NER corpus, the system trained for finding clinical entities was basic. CRFSuite was used with word features and orthographic features. Due to the large number of datasets no hyperparameter tuning was done. The hyperparameter's used are the default parameters for CRFSuite, but with an added $c1$ of 0.1 and a $c1$ of 0.2. 5-fold cross validation was used.

The overlap of classified clinical entities and PHI are investigated, to see how they relate to each other. For *class, mask* and *pseudo* this overlap leads to a

---

[3]Health Bank, `http://dsv.su.se/healthbank`

loss of information and training examples for the clinical NER task. The overlap is analysed on a token level, since one PHI entity may span multiple clinical entities, or only parts of a clinical entity.

In comparison to the other concealment strategies, the de-identification's precision with *Remove* has a clear impact in terms of the overlap between clinical entities and PHI. Sentences with a co-occurrence of at least one token classified as a PHI and an annotated clinical entity are removed with this method. This means that there are fewer examples of clinical entities in the datasets de-identified with *Remove* as the concealment strategy. In the co-occurrence analysis, the co-occurence is measured to get information about how the identification of PHI affects the clinical entity information when using the *Remove* method.

4 Results

The results for PHI identification cross-validated on the PHI corpus are first presented to provide an estimate of the performance of each PHI model. The impact of de-identification on the three clinical NER corpora used for the downstream tasks are then analysed in three different ways: (i) the overlap between predicted PHI and manually annotated clinical entities, (ii) the co-occurrence of predicted PHI and manually annotated clinical entities in the same sentences, and (iii) the impact on downstream clinical NER performance. Finally, the misclassification of eponyms as PHI is specifically studied.

4.1 PHI Identification

The results from the development of PHI models optimised for different F-scores are presented below. The best combination of thresholds for each F-score, F_1, F_4, F_{10}, F_{20} and F_{40}, are presented in Tables 3 and 4 in the Appendix.

The cross-validated performance scores of the PHI identification models with adjusted marginal scores are shown in Table 1. As expected, with increased bias, recall increases at the expense of precision. While the recall improves from *model F_1* to *model F_{40}* by a total of 7 percentage points, the precision drops by as much as 83 percentage points. As expected, the F_1-score is highest when using the model optimised for F_1. The highest number of false positives was observed for the class *Health Care Unit*.

PHI Model	P	R	F_1
F_1	96.07	92.82	94.41
F_4	77.82	97.55	86.57
F_{10}	44.95	99.53	61.93
F_{20}	26.47	99.72	41.83
F_{40}	12.74	99.94	22.60

Table 1: Token-based binary evaluation of the de-identification (PHI) models with differently set marginal scores, optimised for a particular F-score. P stands for precision, R for recall and F_1 for F_1-score.

4.2 Overlap Analysis

Only 1% of clinical entities are affected by the de-identification process for F_1, in terms of partial overlap with PHI entities. As expected, a larger recall bias leads to an increasing amount of tokens being classified as PHI. As many as around one third of all clinical entities are classified as PHI by the F_{40} model. As can be seen in Figure 1, *Health Care Unit*, in comparison to other PHI classes, overlaps more with clinical entities for the models with low precision, while *Person* overlaps more with other PHI in the models with high precision. In relation to the number of classified cases, *Location* is, however, the PHI class that overlaps the most with the clinical entities (26% of all classified locations).

The different clinical entities are affected to different degrees. As can be seen in Figure 1, the clinical entity that overlaps the most with PHI is *Drug*, and the one that overlaps the least with PHI is *Finding*.

4.3 Co-occurrence Analysis

As can be seen in Figure 2, the *Remove* concealment strategy leads to the removal of a large amount of clinical entities. As expected, the PHI identification models with lower precision and higher recall removes more PHI compared to the other models. For F_1, F_4, F_{10} and F_{20}, a greater degree of sentences without clinical entities are removed than with clinical entities. For F_{40}, there is an equal amount removed.

While there is little overlap between the PHI class Age and clinical entities, a disorder is mentioned in 67% of sentences with an age annotation in F_1, for example *82 year old man with Alzheimers*. With the *Remove* concealment strategy, these mentions of disorders will be removed from the de-identified dataset.

Overlap Between Clinical Entities and PHI Entities

	Age	Date	Health care unit	Location	Organisation	Person	Phone Number
F1 Body part	0/169	2/847	1/2698	1/124	0/6	2/1341	0/58
F1 Disorder	0/169	1/847	13/2698	0/124	0/6	21/1341	0/58
F1 Drug	0/169	0/847	11/2698	0/124	0/6	4/1341	0/58
F1 Finding	0/169	0/847	9/2698	0/124	0/6	5/1341	0/58
F4 Body part	0/185	4/1622	29/4617	2/135	0/35	24/1601	0/133
F4 Disorder	0/185	5/1622	104/4617	0/135	0/35	37/1601	0/133
F4 Drug	0/185	6/1622	118/4617	2/135	6/35	39/1601	0/133
F4 Finding	0/185	3/1622	68/4617	2/135	0/35	21/1601	0/133
F10 Body part	0/248	5/2788	151/8882	15/243	5/349	92/2638	4/318
F10 Disorder	0/248	14/2788	341/8882	0/243	4/349	116/2638	0/318
F10 Drug	0/248	22/2788	448/8882	16/243	32/349	229/2638	3/318
F10 Finding	0/248	6/2788	277/8882	3/243	8/349	113/2638	2/318
F20 Body part	0/297	15/3644	416/15829	39/415	15/862	145/4248	7/498
F20 Disorder	0/297	33/3644	677/15829	1/415	14/862	175/4248	2/498
F20 Drug	0/297	27/3644	807/15829	34/415	62/862	468/4248	5/498
F20 Finding	0/297	18/3644	738/15829	8/415	26/862	250/4248	4/498
F40 Body part	0/343	39/5227	957/31837	103/763	67/3089	186/8623	13/1250
F40 Disorder	0/343	49/5227	1472/31837	4/763	43/3089	274/8623	11/1250
F40 Drug	0/343	48/5227	1405/31837	65/763	145/3089	852/8623	12/1250
F40 Finding	4/343	76/5227	2144/31837	23/763	163/3089	579/8623	23/1250

Figure 1: Presentation of clinical entities replaced during de-identification and which label they were classified as. The number shown for each index is the how many overlaps per identified PHI in total, and the colour of each column represents the percentage of clinical entities (see left column) that are overlapped for each clinical entity class.

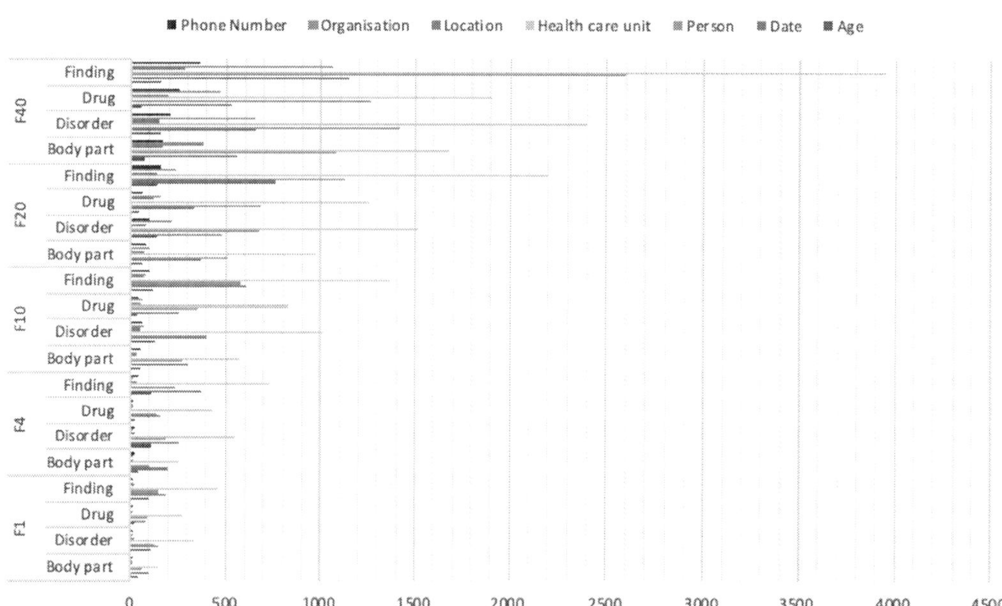

Figure 2: The figure presents the number of sentences with a co-occurrence of a clinical entity and a predicted PHI.

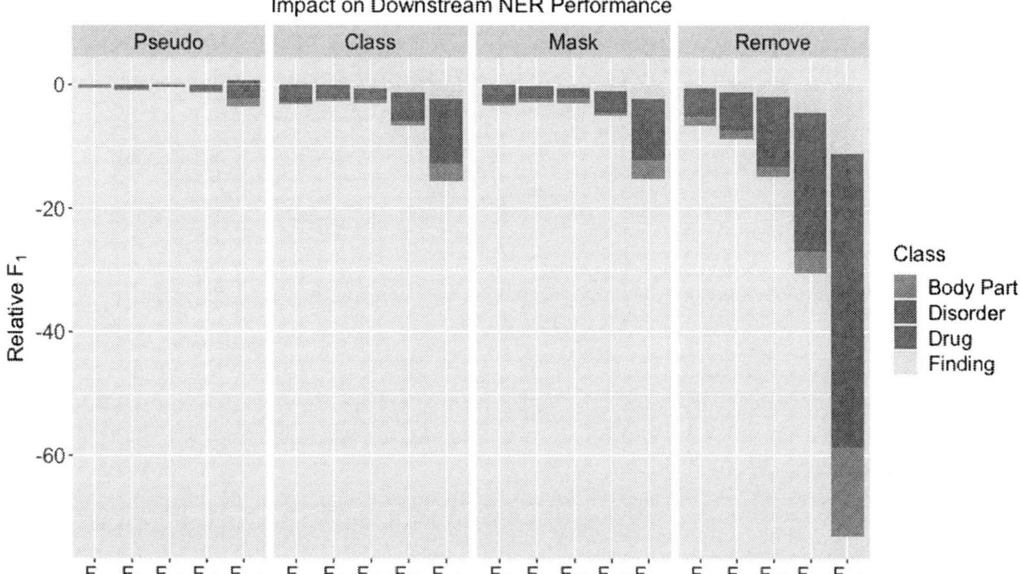

Figure 3: The impact on predictive performance, in terms of relative F_1-score, when training clinical NER models with de-identified data (compared to without de-identification). Results for the target clinical entity classes are shown for five different PHI identification models and four different concealment strategies.

4.4 Downstream NER Performance

As can be seen in Figure 3, the choice of concealment strategy for de-identification has a large impact on downstream clinical NER performance. With the *Pseudo* concealment strategy, the overall performance across models is fairly limited, whereas the impact is somewhat larger with the *Mask* and *Class* concealment strategies, with very little difference between them. A much bigger, negative impact is observed when applying the *Remove* concealment strategy.

The impact on the downstream clinical NER tasks are also different across PHI models. As expected, the performance tends to get increasingly worse the more the PHI model is trained to prioritise recall over precision. However, with *Pseudo*, *Class* and *Mask*, the differences are fairly small for the F_1, F_4 and F_{10} PHI models. With *Remove*, on the other hand, the differences across PHI models are markedly more pronounced.

The impact on performance is not equal across clinical NER classes. Overall – across PHI models and concealment strategies – the most negative impact was observed for the *Drug* class (-3.9%), followed by the *Disorder* class (-3.0%), the *Body Part* class (-1.7%) and the *Finding* class (-1.4%).

Across concealment strategies, the *Drug* class was almost invariably the most impacted clinical entity, with the exception of the F_{20} PHI model, with relative F_1-scores ranging from -1.9% to -9.1%.

The least impacted class varied across PHI models, but was mostly *Body Part* or *Finding*, with relative F_1-scores ranging from -0.4% to -5.4%. In almost all cases, a monotonic decrease in performance is observed as the PHI models are giving increasing priority to recall at the expense of precision. Across PHI models, a similar pattern is observed, with the performance on the *Drug* class being negatively affected the most, ranging from -0.8% to -9.1%. However, with *Remove*, there is a greater impact on the *Disorder* class than the *Drug* class. For all concealment strategies, the *Finding* class is the least affected.

4.5 Eponym Analysis

According to previous studies, medical eponyms risk being mistaken as identifiable information, as they are derived from a personal name.

Throughout the three corpora, 21 different eponyms occurred, with a total of 57 mentions. With 43% of eponyms only being mentioned once, it would not be possible to make any conclusions based on that data. Therefore, an eponym

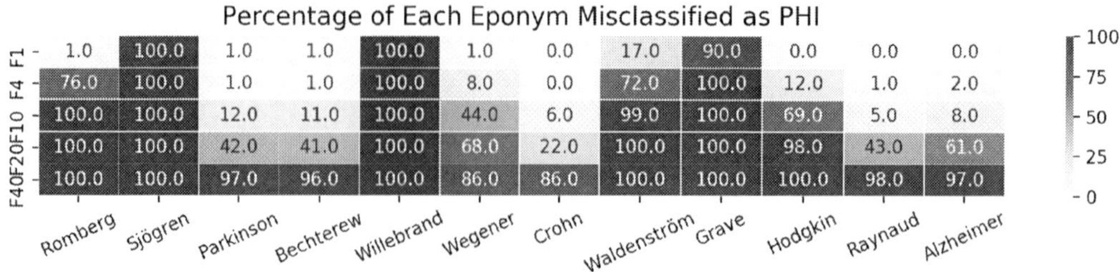

Figure 4: The figure shows how many percent of each eponym that are missclassified as PHI in the Eponym Corpus for different five PHI identification models, F_1, F_4, F_{10}, F_{20} and F_{40}.

dataset was created to be able to study the effect on eponyms further.

In the training data from the PHI corpus, *Parkinson* occurs 11 times, *Bechterew* 4 times, *Waldenström* 2 times, *Alzheimer* 1 time and *Romberg* 1 time. As shown in Figure 4, the eponyms that occurred in the training data are affected to a lesser extent than those that did not. *Sjögren syndrome* and *von Willebrand disease* are regardless of context classified as surnames by all models, together with *Grave's disease. Hodgkin lymphoma* and *Raynaud syndrome*, are never mentioned in the training set, but are classified as surnames less often than for example *Waldenström*. This is despite *Waldenström* being in the training data. With the F_{40} PHI identification model, almost all eponyms are classified as PHI, where 60% are misclassified as *Last Name* and 15% as *First Name*.

To summarise the analysis, the overlap for eponyms with PHI is greater than for disorders in general. The eponyms most likely to be affected by de-identification are those which bear a resemblance to common Swedish last names.

5 Discussion

In contrast to the findings by both Deleger et al. (2013) and Obeid et al. (2019), a small negative impact was indeed observed regardless of the precision and recall of the PHI model used for de-identification. However, the impact is very small unless the F_{40} PHI model or the *Remove* concealment strategy is used.

The worst performing model in (Stubbs et al., 2015) had a token-based binary precision of 76% (and a recall of 52%). The only models with an estimated precision above this in our study is F_1 and F_4. Despite the clear difference in precision between F_1 and F_{10}, F_{10} gives a high recall 99.5% for PHI identification, see Table 1, and the conceal-ment strategies *Pseudo*, *Class* and *Mask* still allow for good downstream results. The trade-off between the level of de-identification and data utility may vary depending on the information sensitivity of the data, and assessments should be made on a case-by-case basis. According to our experiments it is to a certain extent, possible to raise recall at the expense of precision without affecting downstream performance. It is also important to balance this with the chosen concealment method, as they affect the utility to varying degrees. For example, F_{40} may be appropriate to combine with *Pseudo*, but not with *Remove*. Another possibility would be to use different concealment methods for different PHI, since some are not as sensitive as others, and create a recall bias for some labels but not for others.

PHI models with low precision combined with *Pseudo* as concealment strategy seem to have a much smaller impact on downstream tasks. A potential cause may be that not all *Health Care Unit* tokens were affected by the pseudonymisation.

In this study, the de-identification process is performed after manual annotation process of the clinical entity corpora. If the annotation process was performed after the de-identification, and there was more data available than could be annotated, the *Remove* method may have less impact.

Based on this study, certain PHI classes have a higher risk of overlapping with relevant clinical information than others. We have, however, not investigated how the precision and recall of specific PHI labels affect downstream tasks. The analysis of eponyms, where 75% are classified as either *First Name* or *Last Name* by the F_{40} PHI identification model, indicates that certain classes may have a greater impact on downstream tasks than others. The eponym analysis also indicates that, while the downstream impact is small on the clinical NER corpora, other tasks may be affected to a greater extent. As eponyms risk being misclassified as PHI, there may be a need, if the diseases are relevant to the task, to deal with them specifically, for example by adding rules to avoid being unrecognised as disorders.

This study has focused on utility for clinical entity recognition. Future research may investigate how de-identification impacts other downstream tasks. It may also be of interest to have people read through the texts and see if they are readable and possible to use for research in more qualitative research on electronic health records.

6 Conclusion

This study demonstrates that corpora de-identified using PHI models with a moderate to high precision lead to similar performance when used for downstream clinical NER tasks. The impact is, however, affected by both the choice of concealment strategy and the trade-off between precision and recall, in particular when the precision is low.

Optimising the PHI identification model for F_4 gives a relatively lower precision of 77.82% and a higher recall of 97.55%. Compared to using a standard F_1-optimised model for de-identification, this results in higher privacy and a relatively small negative impact on downstream clinical named entity recognition.

This study indicates that it is possible to increase the recall of models for identifying sensitive information without negatively affecting the use of de-identified text data for training models for clinical named entity recognition

Furthermore the overlap analysis showed that with lower precision, there is an increase of overlap between clinical information and automatically labelled PHI. Some PHI labels overlap with clinical entities more than others. Different clinical entities are also more likely to be affected than others, like *Drug*. Eponyms may also risk being misclassified as last names, and studies interested in those may need to take extra precautions to handle those properly.

References

Hanna Berg, Taridzo Chomutare, and Hercules Dalianis. 2019. Building a De-identification System for Real Swedish Clinical Text Using Pseudonymised Clinical Text. In *Proceedings of the Tenth International Workshop on Health Text Mining and Information Analysis (LOUHI 2019)*, pages 118–125.

Hercules Dalianis. 2019. Pseudonymisation of Swedish electronic patient records using a rule-based approach. In *Proceedings of the Workshop on NLP and Pseudonymisation*, pages 16–23, Turku, Finland. Linköping Electronic Press.

Hercules Dalianis and Sumithra Velupillai. 2010. De-identifying Swedish Clinical Text - Refinement of a Gold Standard and Experiments with Conditional Random Fields. *Journal of Biomedical Semantics*, 1:6.

Louise Deleger, Katalin Molnar, Guergana Savova, Fei Xia, Todd Lingren, Qi Li, Keith Marsolo, Anil Jegga, Megan Kaiser, Laura Stoutenborough, and Imre Solti. 2013. Large-scale evaluation of automated clinical note de-identification and its impact on information extraction. *Journal of the American Medical Informatics Association*, 20(1):84–94.

EU. 2014. Article 29 data protection working party, opinion 05/2014 on anonymisation techniques. *EU, https://ec.europa.eu/justice/article-29/documentation/opinion-recommendation/files/2014/wp216_en.pdf*.

Óscar Ferrández, Brett R South, Shuying Shen, F Jeff Friedlin, Matthew H Samore, and Stéphane M Meystre. 2012. Generalizability and comparison of automatic clinical text de-identification methods and resources. In *AMIA Annual Symposium Proceedings*, volume 2012, page 199. American Medical Informatics Association.

Aron Henriksson, Maria Kvist, Hercules Dalianis, and Martin Duneld. 2015. Identifying adverse drug event information in clinical notes with distributional semantic representations of context. *Journal of biomedical informatics*, 57:333–349.

Tochukwu Iwuchukwu, David J DeWitt, AnHai Doan, and Jeffrey F Naughton. 2007. K-anonymization as spatial indexing: Toward scalable and incremental anonymization. In *2007 IEEE 23rd International Conference on Data Engineering*, pages 1414–1416. IEEE.

Maria Kvist, Maria Skeppstedt, Sumithra Velupillai, and Hercules Dalianis. 2011. Modeling human comprehension of Swedish medical records for intelligent access and summarization systems- Future vision, a physician's perspective. In *Proceedings of 9th Scandinavian Conference on Health Informatics, SHI 2011, Oslo, (Eds.) Fensli and Dale, Tapir Academic Press*, pages 31–35.

Stéphane M Meystre, Óscar Ferrández, F Jeffrey Friedlin, Brett R South, Shuying Shen, and Matthew H Samore. 2014. Text de-identification for privacy protection: A study of its impact on clinical text information content. *Journal of biomedical informatics*, 50:142–150.

Stephane M Meystre, F Jeffrey Friedlin, Brett R South, Shuying Shen, and Matthew H Samore. 2010. Automatic De-Identification of Textual Documents in the Electronic Health Record: A Review of Recent Research. *BMC medical research methodology*, 10(1):70.

Jihad S Obeid, Paul M Heider, Erin R Weeda, Andrew J Matuskowitz, Christine M Carr, Kevin Gagnon, Tami Crawford, and Stephane M Meystre. 2019. Impact of de-identification on clinical text classification using traditional and deep learning classifiers. *Studies in Health technology and Informatics*, 264:283.

Maria Skeppstedt, Maria Kvist, Gunnar H Nilsson, and Hercules Dalianis. 2014. Automatic recognition of disorders, findings, pharmaceuticals and body structures from clinical text: An annotation and machine learning study. *Journal of biomedical informatics*, 49:148–158.

Amber Stubbs, Christopher Kotfila, and Özlem Uzuner. 2015. Automated systems for the de-identification of longitudinal clinical narratives: Overview of 2014 i2b2/uthealth shared task track 1. *Journal of biomedical informatics*, 58:S11–S19.

Charles Sutton and Andrew McCallum. 2012. An introduction to Conditional Random Fields. *Foundations and Trends® in Machine Learning*, 4(4):267–373.

Sumithra Velupillai, Hercules Dalianis, Martin Hassel, and Gunnar H Nilsson. 2009. Developing a standard for de-identifying electronic patient records written in Swedish: precision, recall and F-measure in a manual and computerized annotation trial. *International Journal of Medical Informatics*, 78(12):e19–e26.

Rebecka Weegar, Maria Kvist, Karin Sundström, Søren Brunak, and Hercules Dalianis. 2015. Finding cervical cancer symptoms in Swedish clinical text using a machine learning approach and NegEx. In *AMIA Annual Symposium Proceedings*, volume 2015, page 1296. American Medical Informatics Association.

Weiyi Xia, Raymond Heatherly, Xiaofeng Ding, Jiuyong Li, and Bradley A Malin. 2015. Ru policy frontiers for health data de-identification. *Journal of the American Medical Informatics Association*, 22(5):1029–1041.

A   Appendices

A.1   Hyperparameters for the PHI NER

Parameters	Options
Linesearch	**MoreThuente**, StrongBacktracking, Backtracking
Max iterations	50, **100**, 150, 200, 250
Min freq	0, 3, **5**
Period	**5**, 10, 15
Num memories	3, 6, **9**, 12
c1	0.1, **0.05**, 0.01, 0.05, 0.001, 0,0005
c2	0.1, 0.05, **0.01**, 0.05, 0.001, 0,0005
epsilon	**1e-02**, 1,E-03, 1,E-04, 1,E-05, 1,E-06
delta	**1e-02**, 1,E-03, 1,E-04, 1,E-05, 1,E-06
transitions?	True, **False**
states?	**True**, False

Table 2: This is the options used for the hyper-parameter optimisation with random search. The bold ones are the parameters that together produced the best results for the 100 iterations.

A.2   Thresholds based on Grid Search

Main Non-PHI Threshold	Alt PHI Threshold
0.99999, 0.9999, 0.999, 0.99,	0.00001, 0.0001, 0.0005, 0.001,
0.95, 0.90, 0.85, 0.80,	0.005, 0.01, 0.05, 0.1,
0.75, 0.7, 0.6	0.2, 0.3, 0.4

Table 3: This table shows the options for the grid search used for choosing marginal thresholds for the different PHI models.

Optimised	Main Non-PHI Threshold	Alt PHI Threshold
F_1	0.75	0.1
F_4	0.99	0.05
F_{10}	0.999	0.001
F_{20}	0.9999	0.0001
F_{40}	0.99999	0.00001

Table 4: This table shows the best threshold for the marginal probability score for the predicted non-PHI label (Main Non-PHI Threshold) and the threshold for the next most probable label (Alt PHI Threshold) based on a grid search.

Simple Hierarchical Multi-Task Neural End-To-End Entity Linking for Biomedical Text

Maciej Wiatrak, Juha Iso-Sipilä

BenevolentAI

4-8 Maple St, London

W1T 5HD

{maciej.wiatrak, juha.iso-sipila}@benevolent.ai

Abstract

Recognising and linking entities is a crucial first step to many tasks in biomedical text analysis, such as relation extraction and target identification. Traditionally, biomedical entity linking methods rely heavily on heuristic rules and predefined, often domain-specific features. The features try to capture the properties of entities and complex multi-step architectures to detect, and subsequently link entity mentions. We propose a significant simplification to the biomedical entity linking setup that does not rely on any heuristic methods. The system performs all the steps of the entity linking task jointly in either single or two stages. We explore the use of hierarchical multi-task learning, using mention recognition and entity typing tasks as auxiliary tasks. We show that hierarchical multi-task models consistently outperform single-task models when trained tasks are homogeneous. We evaluate the performance of our models on the biomedical entity linking benchmarks using MedMentions and BC5CDR datasets. We achieve state-of-the-art results on the challenging MedMentions dataset, and comparable results on BC5CDR.

1 Introduction & Related Work

The task of identifying and linking mentions of entities to the corresponding knowledge base is a key component of biomedical natural language processing, strongly influencing the overall performance of such systems. The existing biomedical entity linking systems can usually be broken down into two stages: (1) Mention Recognition (MR) where the goal is to recognise the spans of entity mentions in text and (2) Entity Linking (EL, also referred as Entity Normalisation or Standardisation), which given a potential mention, tries to link it to an appropriate type and entity. Often, the entity linking task includes the Entity Typing (ET) and Entity Disambiguation (ED) as separate steps, with the former task aiming to identify the type of the mention, such as *gene*, *protein* or *disease* before passing it to the entity disambiguation stage, which effectively grounds the mention to an appropriate entity.

Widely studied in the general domain, entity linking is particularly challenging for the biomedical text. This is mostly due to the size of the ontology, (here referred to as the knowledge base), high syntactic and semantic overlap between types and entities, the complexity of terms, as well as the lack of availability of annotated text.

Due to these challenges, the majority of the existing methods rely on hand-crafted complex rules and architectures including semi-Markov methods (Leaman and Lu, 2016), approximate dictionary matching (Wang et al., 2019) or use a set of external domain-specific tools with manually curated ontologies (Kim et al., 2019). These methods often include multiple steps, each of these steps carrying over the errors to the subsequent stages. Nevertheless, these tasks are usually interdependent and have been proven to often benefit from a joint objective (Durrett and Klein, 2014). Recently, both in the general and biomedical domain, there has been a steady shift to neural methods to solve EL (Kolitsas et al., 2018; Habibi et al., 2017), leveraging a range of methods including the use of entity embeddings (Yamada et al., 2016), multi-task learning (Mulyar and McInnes, 2020; Khan et al., 2020), and others (Radhakrishnan et al., 2018). There have also been a plethora of mixed methods combining heuristic approaches such as approximate dictionary matching with language models (Loureiro and Jorge, 2020).

This work focuses on multi-task approaches to end-to-end entity linking, which has already been studied in the biomedical domain. These include ones leveraging pre-trained language models (Peng et al., 2020; Crichton et al., 2017; Khan et al., 2020), model dependency (Crichton et al., 2017) and building out a cross-sharing model structure

Proceedings of the 11th International Workshop on Health Text Mining and Information Analysis, pages 12–17

November 20, 2020. ©2020 Association for Computational Linguistics

https://doi.org/10.18653/v1/P17

(Wang et al., 2019). An interesting approach has been proposed by Zhao et al. (2019), where authors established a multi-task deep learning model that trained NER and EL models in parallel, with each task leveraging feedback from the other. A model with a similar setup and architecture to the one here, casting the EL problem as a simple per token classification problem has been outlined by Broscheit (2019). Nevertheless, its application domain, architecture, and training regime strongly differ from the one proposed here.

In this study, we investigate the use of a significantly simpler model, drawing on a set of recent developments in NLP, such as pre-trained language models, hierarchical and multi-task learning to outline a simple, yet effective approach for biomedical end-to-end entity linking. We evaluate our models on three tasks, mention recognition, entity typing, and entity linking, investigating different task setups and architectures on the MedMentions and BioCreative V CDR corpora.

Our contributions are as follows: (1) we propose and evaluate two simple setups using fully neural end-to-end entity linking models for biomedical literature. We treat the problem as a per token classification or per entity classification problem over the entire entity vocabulary. All the steps included in the entity linking task are performed in a single or two steps. (2) We examine the use of mention recognition and entity typing as auxiliary tasks in both multi-task and hierarchical multi-task learning scenario, proving that hierarchical multi-task models outperform single-task models when tasks are homogeneous. (3) We outline the optimal training regime including adapting the loss for the extreme classification problem.

2 Methods

2.1 Tasks

Our main task, which we refer to as **Entity Linking** (EL) aims at classifying each token or a mention to an appropriate entity concept unique identifier (CUI). In order for the mention to be correctly identified, all tokens for the mention need to have the correct golden annotation. If the model has wrongly predicted the token right after or before the entity's golden annotated span, the entity prediction is wrong at the mention-level (Mohan and Li, 2019). For the per entity setup, where the entity representation is derived through mean pooling of all tokens spanning a predicted entity, both the final

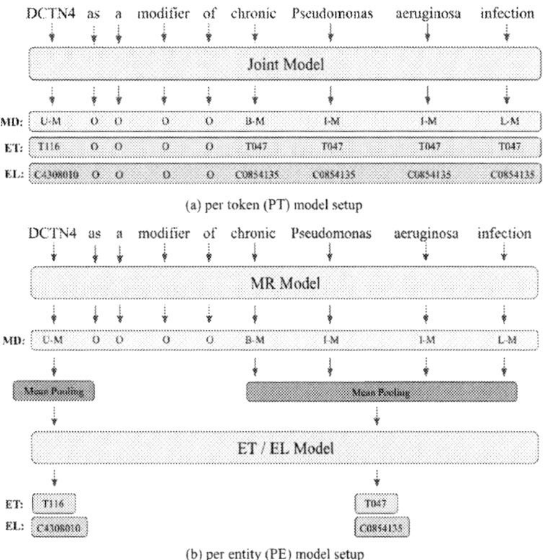

Figure 1: The entity linking setup as a (a) per token (PT) classification and (b) per entity (PE) classification problem with a sentence and corresponding labels for EL, ET and MR, which uses a BILOU scheme for annotations. Here, "O" denotes a *Nil* and "M" a *Mention* prediction.

EL and the MR predictions need to be correct. Figure one provides more information on both setups.

We also make use of two other tasks: **Entity Typing** (ET) and **Mention Recognition** (MR), with the former predicting entity Type Unique Identifier (TUI) for each token and the latter predicting whether a token is a part of the mention. We always use the BILOU scheme for mention recognition token annotation, and due to the low number of types in the BC5CDR dataset, also for the ET task on this corpora. We evaluate the entity prediction at mention-level similarly as in the EL and ET. In per token setup, all three tasks are essentially sequence labelling problems, while in per entity setup, only the MR is a sequence labelling problem and both ET and EL are classification problems leveraging the predictions produced by the MR model.

The reason behind employing ET and MR tasks is for investigating the multi-task learning methods, where we treat ET and MR as auxiliary tasks aimed at regularising and providing additional information to the main EL task leveraging its inherently hierarchical structure. Correspondingly, we also look at the performance impact of the two other tasks on EL task.

2.2 Models

We outline three models: single-task model, multi-task model, and hierarchical multi-task model. The model architecture for the latter two models is depicted on Figure 2. All models take a sentence with

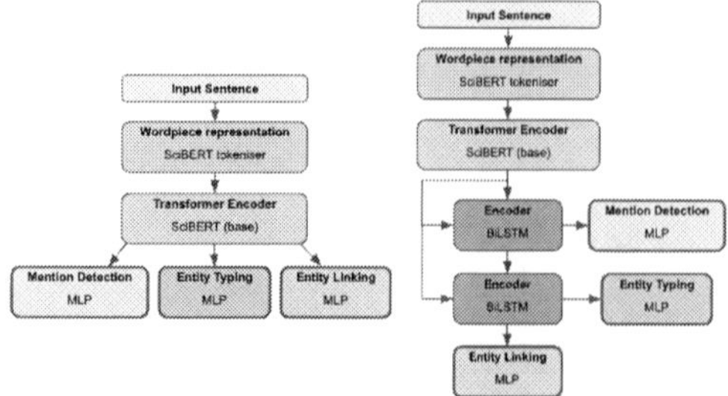

Figure 2: The architectures of the multi-task model (left) and hierarchical multi-task model (right) with hierarchical structure of the tasks and task-specific encoders.

the surrounding context as their input and output a prediction for a token (PT setup) or an average of token embeddings spanning an entity (PE setup). For tokenisation, embedding layer and encoder we use SciBERT (base).

The single-task model only adds a feedforward neural network at the top of the encoder transformer, which acts as a decoder. In the multi-task scenario, three feedforward layers are added on the top of the transformer, each corresponding to a specific task, namely MR, ET, and EL. All of these tasks share the encoder and during a forward pass, the encoder output is fed into each task-specific layers separately, after which the cumulative loss is summed and backpropagated through the model. The intuition behind sharing the encoder is that training on multiple interdependent tasks will act as a regularisation method, thus improving the overall performance and speed of convergence.

The last model is a hierarchical multi-task model that leverages the natural hierarchy between the 3 tasks by introducing an inductive bias by supervising lower level tasks at the bottom layers of the model (MR, ET) and higher level task (EL) at the top layer. Similarly, as in (Sanh et al., 2019), we add task-specific encoders and shortcut connections to process the information from lower to higher level tasks. The higher level tasks take the concatenation of the general transformer encoder output and lower-level task encoder specific output as their input. Here, we use multi-layer BiLSTMs as task-specific encoders.

We experiment with all three models in the per token scenario, as all tasks in this setup are sequence labelling problems. For the per entity framework, we look at a single-task and hierarchical multi-task model, where only the MR step is a sequence labelling task and ET and EL are both classification tasks.

3 Experiments

3.1 Training details

We treat both PE and PT setups as multi-class classification problems over the entire entity vocabulary. In both cases, we use categorical cross-entropy to compute the loss. To address the class imbalance problem in the PT framework, we apply a lower weight to the *Nil* token's output class, keeping other class weights equal. To improve convergence speed and memory efficiency we compute the loss only through the entity classes present in the batch. Therefore, for token t_i in a sequence T, (or correspondingly the mean pooled entity representation from a set of tokens) with a label y_i and its assigned class weight w_k in a minibatch B and entity labels derived from this batch $\hat{E} = E(B)$, the loss is computed by

$$L = -\frac{1}{|B| * |T|} \sum_{k}^{|\hat{E}|} \sum_{j}^{|B|} \sum_{i}^{|T|} w_k y_{ij}^k \log(h_\theta(t_{ij}, k)).$$

Here, y_{ij}^k represents the target label for token i in a sequence j for class k, and $h_\theta(t_{ij}, k)$ represents the model prediction for token t_{ij} and class k, where the parameters θ are defined by the encoder and decoder layers in the model.

We found using the context, namely the sentence after and before the sentence of interest beneficial for the encoder. After encoder, the context sentences are discarded from further steps. For the encoder, we use the SciBERT (base) transformer, and we fine tune the model parameters during training. For the hierarchical multi-task model, we follow the training regime outlined in (Sanh et al., 2019) and found tuning the encoder only on the EL task marginally outperforming sharing it across all three tasks. We treated the *Nil* output class weight as an additional hyperparameter that we set to 0.125 for MedMentions (full) and BC5CDR datasets, and 0.01 for MedMentions st21pv. All trainings were performed using Adam (Kingma and Ba, 2015) with $1e - 4$ weight decay, $2 - e5$ learning rate, batch size of 32 and max sequence length of 128.

Dataset	#Docs	#Mentions	#Unq TUI	#Unq CUI
MedMentions (full)	4,392	352,496	126	34,724
MedMentions (st21pv)	4,392	203,282	21	25,419
Bio CDR	1,500	28,559	2	5,818

Table 1: Details of biomedical entity linking datasets used in our experiments.

	MedMentions(full)				MedMentions (st21pv)				BC5CDR			
	Mention Recognition											
Model	Acc	P	R	F1	Acc	P	R	F1	Acc	P	R	F1
SciSpacy	N/A	69.61	68.56	69.08	N/A	41.23	70.57	52.05	N/A	81.47	73.47	77.81
BiLSTM-CRF	82.47	64.09	65.03	64.56	84.86	60.53	61.7	61.11	94.00	72.09	78.65	75.23
Single-task	85.56	73.4	69.38	**71.33**	87.72	73.55	**66.92**	**70.05**	**97.04**	89.64	**88.25**	**88.94**
Multi-task	**85.62**	72.62	**69.72**	71.14	**87.84**	73.34	66.53	69.76	96.93	**90.56**	87.24	88.87
Hier. Multi-task	85.40	**73.13**	68.93	70.97	85.59	**74.19**	59.25	65.88	96.68	89.31	84.91	87.05

	MedMentions(full)				MedMentions (st21pv)				BC5CDR			
	Entity Typing											
Model	Acc	P	R	F1	Acc	P	R	F1	Acc	P	R	F1
SciSpacy	N/A	39.67	39.08	39.37	N/A	10.14	31.68	15.26	N/A	N/A	N/A	N/A
BiLSTM-CRF	72.26	45.14	44.98	45.06	82.46	47.15	52.29	49.59	94.03	72.08	78.70	75.24
PT-Single-task	78.27	55.79	51.65	53.64	86.67	63.10	58.26	60.59	**96.96**	89.52	87.48	88.45
PE-Single-task	N/A	**57.5**	52.62	**54.95**	N/A	**65.05**	**60.43**	**62.65**	N/A	**90.53**	**87.65**	**89.07**
PT-Multi-task	**78.3**	55.39	**52.66**	53.99	**86.72**	63.77	58.86	61.21	96.90	90.33	87.04	88.65
PT-Hier. Multi-task	76.7	61.94	49.41	50.61	80.87	46.22	40.76	43.32	96.57	88.40	84.24	86.27
PE-Hier. Multi-task	N/A	50.91	46.49	48.65	N/A	59.44	55.27	57.30	N/A	88.15	85.34	86.72

	MedMentions(full)				MedMentions (st21pv)				BC5CDR			
	Entity Linking											
Model	Acc	P	R	F1	Acc	P	R	F1	Acc	P	R	F1
SciSpacy	N/A	34.14	33.63	33.88	N/A	25.17	53.52	34.24	N/A	58.43	52.70	55.42
BiLSTM-CRF*	62.73	39.89	30.25	32.22	71.35	33.65	25.46	28.99	89.52	52.72	47.59	50.02
PT-Single-task	67.98	46.41	39.46	42.65	75.57	44.09	35.58	39.36	91.62	64.14	57.56	60.67
PE-Single-task	N/A	46.3	**42.37**	**44.25**	N/A	43.03	39.97	41.45	N/A	**64.98**	**62.91**	**63.93**
PT-Multi-task	**68.23**	45.88	40.13	42.81	**76.43**	44.03	37.85	40.71	91.45	63.51	54.35	58.62
PT-Hier. Multi-task	68.13	**46.89**	39.93	43.13	76.14	**44.32**	37.69	40.74	**91.65**	64.35	59.27	63.15
PE-Hier. Multi-task	N/A	46.21	42.29	44.16	N/A	43.12	**40.06**	**41.53**	N/A	64.54	62.49	63.5

Table 2: Results: performance of various models on MR, EL and ET tasks on the test sets. Here *Acc-pt* denotes per token accuracy. * for EL task on MedMentions full and st21pv we used a MLP layer on top of BiLSTM instead of CRF due to the lower performance of CRF on large number of output classes.

The models were trained on a single NVIDIA V100 GPU until convergence.

3.2 Datasets and Evaluation metrics

We evaluate our models on three datasets; two versions of the recently released MedMentions dataset; (1) full set and (2) and st21pv subset of it (Mohan and Li, 2019) and BioCreative V CDR task corpus (Li et al., 2016). Each mention in the dataset is labelled with a concept unique identifier (CUI) and type unique identifier (TUI). Both MedMentions datasets target UMLS ontology but vary in terms of number of types and mentions, while the BioCreative V corpora is normalised with MeSH identifiers. The datasets details are summarised in Table 1.

We measure the performance of each task using mention-level metrics described in (Mohan and Li, 2019), providing precision, recall, and F1 scores. Additionally, we record the per token accuracy for the per token setup. As benchmarks, we use SciSpacy (Neumann et al., 2019) package, which has been shown to outperform other biomedical text

processing tools such as QuickUMLS or MetaMap on full MedMentions and BC5CDR (Vashishth et al., 2020). Due to little results reported on the end-to-end entity linking task on MedMentions, we also use BiLSTM-CRF in per token setup as a benchmark.

3.3 Results and discussion

In Tables 2 and 3 we outline the results on MR, ET, and EL tasks. While the reported results are all optimal for single-task models, it should be noted that all multi-task models optimise for the EL task with MR and ET serving as auxiliary tasks, hence the EL is the focus of the discussion. All of the models outlined here significantly outperform SciSpacy and BiLSTM-CRF, particularly in ET and EL. The per entity setup proves to perform better on EL than the simpler per token framework by 0.87 F1 points on average, yielding particularly better recall results (2.03 points). Error analysis has shown that this is often due to the lexical overlap of some *Nil* tokens with entity tokens, resulting in a model often assigning an entity label for tokens with gold *Nil* token label. Furthermore, in the per token setup, the multi-task models consistently outperform the single-task models on EL, with the hierarchical multi-task model achieving the best results (on average 1.45 F1 points better than single-task models). In contrast, this has not been the case for the per entity framework, where the single-task models have on average performed marginally better on EL. We hypothesise that this is due to the homogeneity of the tasks in the per token setup, with all the tasks being sequence labelling problems, which is not the case for the per entity case. Interestingly, the achieved results are higher for the full MedMentions dataset than for the st21pv subset. This highlights the problem of achieving high macro performance mentioned in (Loureiro and Jorge, 2020) for biomedical entity linking.

4   Conclusion & Future Work

In this work, we have proposed a simple neural approach to end-to-end entity linking for biomedical text which makes no use of heuristic features. We have proven that the problem can benefit from the hierarchical multi-task learning when tasks are homogeneous. We report state-of-the-art results on EL on the full MedMentions dataset and comparable results on the MR and ET tasks on BC5CDR (Zhao et al., 2019). The work could easily be extended by, for example, using the output of the PT setup as features or by further developing the hierarchical multi-task framework of end-to-end entity linking problem. Moreover, the additional parameters such as output class weights or loss scaling which has not been used here could be easily adapted to a particular problem.

Acknowledgments

We thank Theodosia Togia, Felix Kruger, Mikko Vilenius and Jonas Vetterle for helpful feedbacks and the anonymous reviewers for constructive comments on the paper.

References

Samuel Broscheit. 2019. Investigating entity knowledge in BERT with simple neural end-to-end entity linking. In *Proceedings of the 23rd Conference on Computational Natural Language Learning (CoNLL)*, pages 677–685, Hong Kong, China. Association for Computational Linguistics.

Gamal Crichton, Sampo Pyysalo, Billy Chiu, and Anna Korhonen. 2017. A neural network multi-task learning approach to biomedical named entity recognition. *BMC Bioinformatics*, 18(1):1–14.

Greg Durrett and Dan Klein. 2014. A Joint Model for Entity Analysis: Coreference, Typing, and Linking. *Transactions of the Association for Computational Linguistics*, 2:477–490.

Maryam Habibi, Leon Weber, Mariana Neves, David Luis Wiegandt, and Ulf Leser. 2017. Deep learning with word embeddings improves biomedical named entity recognition. *Bioinformatics*, 33(14):i37–i48.

Muhammad Raza Khan, Morteza Ziyadi, and Mohamed Abdelhady. 2020. MT-BioNER: Multi-task Learning for Biomedical Named Entity Recognition using Deep Bidirectional Transformers. *ArXiv*, abs/2001.08904.

Donghyeon Kim, Jinhyuk Lee, Chan H O So, Hwisang Jeon, Minbyul Jeong, Yonghwa Choi, Wonjin Yoon, Mujeen Sung, and Jaewoo Kang. 2019. A Neural Named Entity Recognition and Multi-Type Normalization Tool for Biomedical Text Mining. *IEEE Access*, 7:73729–73740.

Diederik P. Kingma and Jimmy Ba. 2015. Adam: A method for stochastic optimization. In *3rd International Conference on Learning Representations, ICLR 2015, San Diego, CA, USA, May 7-9, 2015, Conference Track Proceedings*.

Nikolaos Kolitsas, Octavian-Eugen Ganea, and Thomas Hofmann. 2018. End-to-End Neural Entity Linking. *ArXiv*, abs/1808.07699.

Robert Leaman and Zhiyong Lu. 2016. TaggerOne: joint named entity recognition and normalization with semi-Markov Models. *Bioinformatics*, 32(18):2839–2846.

Jiao Li, Yueping Sun, Robin J. Johnson, Daniela Sciaky, Chih-Hsuan Wei, Robert Leaman, Allan Peter Davis, Carolyn J. Mattingly, Thomas C. Wiegers, and Zhiyong Lu. 2016. BioCreative V CDR task corpus: a resource for chemical disease relation extraction. *Database*, 2016. Baw068.

Daniel Loureiro and Alípio Jorge. 2020. MedLinker: Medical Entity Linking with Neural Representations and Dictionary Matching. *Advances in Information Retrieval*, 12036:230 – 237.

Sunil Mohan and Donghui Li. 2019. MedMentions: A Large Biomedical Corpus Annotated with UMLS Concepts. In *In Proceedings of the 2019 Conference on Automated Knowledge Base Construction (AKBC 2019). Amherst, Massachusetts, USA. May 2019*.

Andriy Mulyar and Bridget T. McInnes. 2020. MT-Clinical BERT: Scaling Clinical Information Extraction with Multitask Learning.

Mark Neumann, Daniel King, Iz Beltagy, and Waleed Ammar. 2019. ScispaCy: Fast and Robust Models for Biomedical Natural Language Processing. In *Proceedings of the 18th BioNLP Workshop and Shared Task*, pages 319–327, Florence, Italy. Association for Computational Linguistics.

Yifan Peng, Qingyu Chen, and Zhiyong Lu. 2020. An empirical study of multi-task learning on bert for biomedical text mining. In *In BioNLP 2020 Workshop on Biomedical Natural Language Processing*.

Priya Radhakrishnan, Partha Talukdar, and Vasudeva Varma. 2018. ELDEN: Improved entity linking using densified knowledge graphs. In *Proceedings of the 2018 Conference of the North American Chapter of the Association for Computational Linguistics: Human Language Technologies, Volume 1 (Long Papers)*, pages 1844–1853, New Orleans, Louisiana. Association for Computational Linguistics.

Victor Sanh, Thomas Wolf, and Sebastian Ruder. 2019. A hierarchical multi-task approach for learning embeddings from semantic tasks. In *AAAI*.

Shikhar Vashishth, Rishabh Joshi, Ritam Dutt, Denis Newman-Griffis, and Carolyn Rose. 2020. MedType: Improving Medical Entity Linking with Semantic Type Prediction.

Xi Wang, Jiagao Lyu, Li Dong, and Ke Xu. 2019. Multitask learning for biomedical named entity recognition with cross-sharing structure. *BMC Bioinformatics*, 20(1):1–13.

Ikuya Yamada, Hiroyuki Shindo, Hideaki Takeda, and Yoshiyasu Takefuji. 2016. Joint learning of the embedding of words and entities for named entity disambiguation. In *Proceedings of The 20th SIGNLL Conference on Computational Natural Language Learning*, pages 250–259, Berlin, Germany. Association for Computational Linguistics.

Sendong Zhao, Ting Liu, Sicheng Zhao, and Fei Wang. 2019. A Neural Multi-Task Learning Framework to Jointly Model Medical Named Entity Recognition and Normalization. In *AAAI*.

Medical Concept Normalization in User-Generated Texts by Learning Target Concept Embeddings

Katikapalli Subramanyam Kalyan
Department of Computer Applications
NIT Trichy, India
kalyan.ks@yahoo.com

Sivanesan Sangeetha
Department of Computer Applications
NIT Trichy, India
sangeetha@nitt.edu

Abstract

Medical concept normalization helps in discovering standard concepts in free-form text i.e., maps health-related mentions to standard concepts in a clinical knowledge base. It is much beyond simple string matching and requires a deep semantic understanding of concept mentions. Recent research approach concept normalization as either text classification or text similarity. The main drawback in existing a) text classification approach is ignoring valuable target concepts information in learning input concept mention representation b) text similarity approach is the need to separately generate target concept embeddings which is time and resource consuming. Our proposed model overcomes these drawbacks by jointly learning the representations of input concept mention and target concepts. First, we learn input concept mention representation using RoBERTa. Second, we find cosine similarity between embeddings of input concept mention and all the target concepts. Here, embeddings of target concepts are randomly initialized and then updated during training. Finally, the target concept with maximum cosine similarity is assigned to the input concept mention. Our model surpasses all the existing methods across three standard datasets by improving accuracy up to 2.31%.

1 Background

Internet users use social media to voice their views and opinions. Medical social media is a part of social media in which the focus is limited to health and related issues (Pattisapu et al., 2017). User generated texts in medical social media include tweets, blog posts, reviews on drugs, health related question and answers in discussion forums. This rich source of data can be utilized in many health related applications to enhance the quality of services provided (Kalyan and Sangeetha, 2020b).

Medical concept normalization aims at discovering standard medical concepts in free-form text. In this task, health related mentions are mapped to standard concepts in a clinical knowledge base. For example, the concept mention *'hard to stay awake'* is mapped to the standard concept *'drowsy'*. The common public express their health related conditions in an informal way using layman terms while clinical knowledge base contains concepts expressed in scientific language. This variation (colloquial vs scientific) in the languages of common public and knowledge bases makes concept normalization an essential step in understanding user-generated texts. This task is much beyond simple string matching as the same concept can be expressed in a descriptive way using colloquial words or in multiple ways using aliases, acronyms, partial names and morphological variants. Further, noisy nature of user-generated texts and the short length of health-related mentions make the task of concept normalization more challenging.

Research in medical concept normalization started with string matching techniques (Aronson, 2001; McCallum et al., 2005; Tsuruoka et al., 2007) followed by machine learning techniques (Leaman et al., 2013; Leaman and Lu, 2014). The inability of these methods to consider semantics into account shifted research towards deep learning methods with embeddings as input (Limsopatham and Collier, 2016; Lee et al., 2017; Tutubalina et al., 2018; Subramanyam and Sangeetha, 2020). For example, Lee et al. (2017) and Tutubalina et al. (2018) experimented with RNN on the top of domain specific embeddings. Further, lack of large labeled datasets and necessity to train deep learning models like CNN or RNN from scratch (except embeddings) shifted research towards using pretrained language models like BERT and RoBERTa (Miftahutdinov and Tutubalina, 2019; Kalyan and Sangeetha, 2020a; Pattisapu et al., 2020). Miftahut-

Proceedings of the 11th International Workshop on Health Text Mining and Information Analysis, pages 18–23
November 20, 2020. ©2020 Association for Computational Linguistics
https://doi.org/10.18653/v1/P17

dinov and Tutubalina (2019) experimented with BERT based fine-tuned models while Kalyan and Sangeetha (2020a) provided a comprehensive evaluation of BERT based general and domain specific models. The approach of Pattisapu et al. (2020) is based on RoBERTa (Liu et al., 2019) and graph embedding based target concept vectors. The main drawbacks in existing work are :

- text classification approach (Limsopatham and Collier, 2016; Lee et al., 2017; Subramanyam and Sangeetha, 2020; Kalyan and Sangeetha, 2020a) is not exploiting target concepts information in learning input concept mention representation . However, recent work in various natural language processing and computer vision tasks highlights the importance of exploiting target label information in learning input representation. (Rodriguez-Serrano et al., 2013; Akata et al., 2015; Wang et al., 2018; Pappas and Henderson, 2019; Liu et al., 2020).

- text similarity approach of Pattisapu et al. (2020) is the need to generate target concept embeddings separately using graph embedding methods. This is time and resource consuming when different vocabularies are used for mapping in different data sets (e.g., SNOMED-CT is used in CADEC (Karimi et al., 2015) and PsyTAR (Zolnoori et al., 2019) datasets, MedDRA (Mozzicato, 2009) is used in SMM4H2017 (Sarker et al., 2018)). Moreover, the quality of generated concept embeddings using graph embedding methods depends on the comprehensiveness of vocabulary. For example, MedDRA is less fine grained compared to SNOMED-CT (Bodenreider, 2009). This requirement of comprehensive vocabulary limits the effectiveness of this approach.

Our model normalizes input concept mention by jointly learning the representations of input concept mention and target concepts. By learning the representations of target concepts along with input concept mention, our model a) exploits target concepts information unlike existing text classification approaches (Tutubalina et al., 2018; Miftahutdinov and Tutubalina, 2019; Kalyan and Sangeetha, 2020a) and b) eliminates the time and resource consuming process of separately generating target concept embeddings unlike existing text similarity

approach (Pattisapu et al., 2020). Our key contributions are :

- We propose a simple and novel approach which exploits the target concepts information in normalizing concept mention by jointly learning the representations of input concept mention and all the target concepts. It is the first work in medical concept normalization which jointly learns the representations of input concept mention and the target concepts.

- Our model achieves the best results across three standard data sets surpassing all the existing methods with an accuracy improvement of up to 2.31%.

2 Methodology

2.1 Model Description

Our model normalizes concept mentions in two phases. First, it learns input concept mention representation using RoBERTa (Liu et al., 2019). Second, it finds cosine similarity between embeddings of input concept mention and all the target concepts. Here, embeddings of target concepts are randomly initialized and then updated during training. Finally, the target concept with maximum cosine similarity is assigned to the input concept mention.

Input concept mention is encoded into a fixed size vector $m \in \mathbb{R}^d$ using RoBERTa. RoBERTa is a contextualized embedding model pre-trained on 160 GB of text corpus. It consists of an embedding layer followed by a sequence of transformer encoders (Liu et al., 2019).

$$m = RoBERTa(mention) \qquad (1)$$

Input concept mention vector m is transformed into cosine similarity vector $q \in \mathbb{R}^N$ by finding cosine similarity between m and randomly initialized embeddings $\{c_1, c_2, c_3, \ldots c_N\}$ of all target concepts $\{C_1, C_2, \ldots C_N\}$ where $c_i \in \mathbb{R}^d$ and N represents total number of unique target concepts in the dataset. During training, the target concept embeddings and parameters of RoBERTa are updated. Here d is equal to size of hidden state vector in RoBERTa (768 in RoBERTa-base and 1024 in RoBERTa-large).

$$q = [q_i]_{i=1}^N \; where \; q_i = CS(m, c_i) \qquad (2)$$

Here $i = 1, 2, 3, \ldots N$ and the function $CS()$ represents cosine similarity defined as

$$CS(\boldsymbol{m}, \boldsymbol{c}) = \frac{\sum_{i=1}^{d} m_i \times c_i}{\sqrt{\sum_{i=1}^{d} (m_i)^2} \times \sqrt{\sum_{i=1}^{d} (c_i)^2}} \quad (3)$$

Cosine similarity vector $\boldsymbol{q}$ is normalized to $\hat{\boldsymbol{q}}$ using softmax function.

$$\hat{\boldsymbol{q}} = Softmax(\boldsymbol{q}) \quad (4)$$

Finally, model parameters and target concept embeddings are updated using AdamW optimizer (Loshchilov and Hutter, 2019) which minimizes cross entropy loss ($\mathcal{L}$) between normalized cosine similarity vector $\hat{\boldsymbol{q}}$ and one hot encoded ground truth vector $\boldsymbol{p} \in \mathbb{R}^N$. Here M represents number of training instances.

$$\mathcal{L} = -\frac{1}{M} \sum_{i=1}^{M} \sum_{j=1}^{N} p_j^i log(\hat{q}_j^i) \quad (5)$$

2.2 Evaluation Metric

We evaluate our normalization system using accuracy metric, as in the previous works (Miftahutdinov and Tutubalina, 2019; Kalyan and Sangeetha, 2020a; Pattisapu et al., 2020). Accuracy represents the percentage of correctly normalized mentions. In case of CADEC (Karimi et al., 2015) and PsyTAR (Zolnoori et al., 2019) datasets which are multi-fold, reported accuracy is average accuracy across folds.

3 Experimental Setup

3.1 Implementation Details

Pre-processing of input concept mentions include a) removal of non-ASCII and special characters b) normalizing words with more than two consecutive repeating characters (e.g., sleeep $\rightarrow$ sleep) and c) replacing English contraction and medical acronym words with corresponding full forms (e.g., can't $\rightarrow$ cannot, bp $\rightarrow$ blood pressure). The list of medical acronyms is gathered from acronymslist.com and Wikipedia. Pattisapu et al. (2020) generate additional labeled instances by considering synonyms in mapping lexicon as user-geneated concept mentions and augment training set with these labeled instances. However, we don't augment the training set with any additional labeled instances generated from mapping lexicon and we use only the training instances available in the datasets . We choose 10%

of training set for validation and find optimal hyperparameter values using random search. We use AdamW optimizer (Loshchilov and Hutter, 2019) with a learning rate of 3e-5. The final results reported are based on the optimal hyperparameter settings. To implement our model, we choose PyTorch framework and transformers library (Wolf et al., 2019).

3.2 Datasets

SMM4H2017 : This dataset is released for task3 of SMM4H 2017 (Sarker et al., 2018) shared tasks. It consists of ADR phrases extracted from twitter using drug names as keywords and then mapped to Preferred Terms (PTs) from MedDRA. In this, training set includes 6650 phrases assigned with 472 PTs and test set includes 2500 phrases assigned with 254 PTs.

CADEC: CSIRO Adverse Drug Event Corpus (CADEC) includes user generated medical reviews related to Diclofenac and Lipitor (Karimi et al., 2015). The manually identified health related mentions are mapped to target concepts in SNOMED-CT vocabulary. The dataset includes 6,754 mentions mapped to one of the 1029 SNOMED-CT codes. As the random folds of CADEC dataset created by Limsopatham and Collier (2016) have significant overlap between train and test instances, Tutubalina et al. (2018) create custom folds [1] of this dataset with minimum overlap.

PsyTAR: Psychiatric Treatment Adverse Reactions (PsyTAR) corpus includes psychiatric drug reviews obtained from AskaPatient (Zolnoori et al., 2019). Zolnoori et al. (2019) manually identify 6556 health related mentions and map them to one of 618 SNOMED-CT codes. Due to significant overlap between train and test sets of random folds released by Zolnoori et al. (2019), Miftahutdinov and Tutubalina (2019) create custom folds[2] of this dataset with minimum overlap.

We evaluate our model using SMM4H2017, custom folds of CADEC and PsyTAR datasets.

4 Results

Table 1 provides a comparison of our model and the existing methods across three standard concept normalization datasets CADEC, PsyTAR and

---

[1] https://cutt.ly/Gi6kka6
[2] https://doi.org/10.5281/zenodo.3236318

Method	CADEC	PsyTAR	SMM4H17
(Tutubalina et al., 2018)	70.05	-	-
(Subramanyam and Sangeetha, 2020)	75.12	-	-
(Han et al., 2017)	-	-	87.20
(Belousov et al., 2017)	-	-	87.70
(Miftahutdinov and Tutubalina, 2019)	79.83	77.52	89.64
(Kalyan and Sangeetha, 2020a)	82.62	-	-
(Pattisapu et al., 2020)	83.18	82.42	-
Roberta-base + concept embeddings$^{\perp}$	82.60	81.90	90.15
Roberta-large + concept embeddings$^{\perp}$	**85.49 (2.31 ↑)**	**83.68 (1.26 ↑)**	**90.84 (1.2 ↑)**

Table 1: Accuracy of existing methods and our proposed model across CADEC, PsyTAR and SMM4H2017 datasets. $\perp$ - concept embeddings are randomly initialized and then updated during training.

SMM4H2017. The first seven rows represent existing systems and the next two rows represent our approach. Our model achieves new state-of-the-art accuracy of 85.49%, 83.68% and 90.84% across three datasets. Our model outperforms the existing state-of-the-art method of Pattisapu et al. (2020) with accuracy improvement of 2.31%, 1.26% and 1.2% respectively. We didn't augment the training set with labeled instances generated out of synonyms from mapping lexicon like Pattisapu et al. (2020), but still our approach achieved significant improvements. State-of-the-art results achieved by our model across three standard datasets illustrate that learning target concept representations along with input mention representations is simple and much effective compared to separately generating target concept representations using graph embedding methods and then using them.

5 Analysis

Here, we discuss merits and demerits of our proposed method.

5.1 Merit Analysis

We illustrate the effectiveness of our approach in the following two cases.

- In case I, existing methods map the concept mention '*no concentration*' to a closely related target concept '*Poor concentration (26329005)*' instead of the correct target concept '*Unable to concentrate (60032008)*'. Similarly, '*sleepy*' is mapped to '*hypersomnia (77692006)*' instead of '*drowsy (271782001)*'.

- In case II, '*horrible pain*' is mapped to abstract target concept '*Pain (22253000)*' instead of fine-grained target concept '*Severe pain (76948002)*'. Similarly, '*fatigue in arms*' is mapped to '*fatigue (84229001)*' instead of '*muscle fatigue (80449002)*'.

In both the cases, existing methods are unable to exploit target concept information effectively and fail to assign the correct concept. However, our approach exploits target concept information by jointly learning representations of input concept mention and target concepts and hence assigns the concepts correctly.

5.2 Demerit Analysis

Our model aims to map health related mentions to standard concepts. We observe the predictions of our model and identify the following errors.

- In case I, errors are related to insufficient number of training instances. For example, '*hard to stay awake*' is assigned with more frequent concept '*insomnia (193462001)*' instead of the ground truth concept '*drowsy (271782001)*'. Similarly '*muscle cramps in lower legs*' is assigned with '*cramp in lower limb (449917004)*' instead of '*cramp in lower leg (449918009)*'.

- In case II, errors are related to the inability in learning appropriate representations for domain specific rare words. For example, the mentions '*pruritus*' and '*hematuria*' are assigned to completely unrelated concepts '*Tinnitus (60862001)*' and '*diarrhea (62315008)*' respectively.

6 Conclusion

In this work, we deal with medical concept normalization in user generated texts. Our model overcomes the drawbacks in existing text classification and text similarity approaches by jointly learning the representations of input concept mention and target concepts. By learning target concept representations along with input concept mention representations, our approach a) exploits valuable target concepts information unlike existing text classification approaches and b) eliminates the need to separately generate target concept embeddings unlike existing text similarity approach. Our model surpasses all the existing methods across three standard datasets by improving accuracy up to 2.31%. In future, we would like to explore other possible options to include target concept information which may further improve the results.

References

Zeynep Akata, Florent Perronnin, Zaid Harchaoui, and Cordelia Schmid. 2015. Label-embedding for image classification. *IEEE transactions on pattern analysis and machine intelligence*, 38(7):1425–1438.

Alan R Aronson. 2001. Effective mapping of biomedical text to the umls metathesaurus: the metamap program. In *Proceedings of the AMIA Symposium*, page 17. American Medical Informatics Association.

Maksim Belousov, William Dixon, and Goran Nenadic. 2017. Using an ensemble of generalised linear and deep learning models in the smm4h 2017 medical concept normalisation task.

Olivier Bodenreider. 2009. Using snomed ct in combination with meddra for reporting signal detection and adverse drug reactions reporting. In *AMIA Annual Symposium Proceedings*, volume 2009, page 45. American Medical Informatics Association.

Sifei Han, Tung Tran, Anthony Rios, and Ramakanth Kavuluru. 2017. Team uknlp: Detecting adrs, classifying medication intake messages, and normalizing adr mentions on twitter. In *SMM4H@ AMIA*, pages 49–53.

Katikapalli Subramanyam Kalyan and S Sangeetha. 2020a. Bertmcn: Mapping colloquial phrases to standard medical concepts using bert and highway network. Technical report, EasyChair.

Katikapalli Subramanyam Kalyan and S Sangeetha. 2020b. SECNLP: A survey of embeddings in clinical natural language processing. *Journal of biomedical informatics*, 101:103323.

Sarvnaz Karimi, Alejandro Metke-Jimenez, Madonna Kemp, and Chen Wang. 2015. Cadec: A corpus of adverse drug event annotations. *Journal of biomedical informatics*, 55:73–81.

Robert Leaman, Rezarta Islamaj Doğan, and Zhiyong Lu. 2013. Dnorm: disease name normalization with pairwise learning to rank. *Bioinformatics*, 29(22):2909–2917.

Robert Leaman and Zhiyong Lu. 2014. Automated disease normalization with low rank approximations. In *Proceedings of BioNLP 2014*, pages 24–28.

Kathy Lee, Sadid A Hasan, Oladimeji Farri, Alok Choudhary, and Ankit Agrawal. 2017. Medical concept normalization for online user-generated texts. In *2017 IEEE International Conference on Healthcare Informatics (ICHI)*, pages 462–469. IEEE.

Nut Limsopatham and Nigel Collier. 2016. Normalising medical concepts in social media texts by learning semantic representation. Association for Computational Linguistics.

Xien Liu, Song Wang, Xiao Zhang, Xinxin You, Ji Wu, and Dejing Dou. 2020. Label-guided learning for text classification. *arXiv preprint arXiv:2002.10772*.

Yinhan Liu, Myle Ott, Naman Goyal, Jingfei Du, Mandar Joshi, Danqi Chen, Omer Levy, Mike Lewis, Luke Zettlemoyer, and Veselin Stoyanov. 2019. Roberta: A robustly optimized bert pretraining approach. *arXiv preprint arXiv:1907.11692*.

Ilya Loshchilov and Frank Hutter. 2019. Decoupled weight decay regularization. In *International Conference on Learning Representations*.

Andrew McCallum, Kedar Bellare, and Fernando Pereira. 2005. A conditional random field for discriminatively-trained finite-state string edit distance. In *Proceedings of the Twenty-First Conference on Uncertainty in Artificial Intelligence*, pages 388–395.

Zulfat Miftahutdinov and Elena Tutubalina. 2019. Deep neural models for medical concept normalization in user-generated texts. In *Proceedings of the 57th Annual Meeting of the Association for Computational Linguistics: Student Research Workshop*, pages 393–399.

Patricia Mozzicato. 2009. Meddra. *Pharmaceutical Medicine*, 23(2):65–75.

Nikolaos Pappas and James Henderson. 2019. Gile: A generalized input-label embedding for text classification. *Transactions of the Association for Computational Linguistics*, 7:139–155.

Nikhil Pattisapu, Manish Gupta, Ponnurangam Kumaraguru, and Vasudeva Varma. 2017. Medical persona classification in social media. In *Proceedings of the 2017 IEEE/ACM International Conference on Advances in Social Networks Analysis and Mining 2017*, pages 377–384.

Nikhil Pattisapu, Sangameshwar Patil, Girish Palshikar, and Vasudeva Varma. 2020. Medical Concept Normalization by Encoding Target Knowledge. In *Proceedings of the Machine Learning for Health NeurIPS Workshop*, volume 116 of *Proceedings of Machine Learning Research*, pages 246–259. PMLR.

Jose A Rodriguez-Serrano, Florent Perronnin, and France Meylan. 2013. Label embedding for text recognition. In *BMVC*, pages 5–1.

Abeed Sarker, Maksim Belousov, Jasper Friedrichs, Kai Hakala, Svetlana Kiritchenko, Farrokh Mehryary, Sifei Han, Tung Tran, Anthony Rios, Ramakanth Kavuluru, et al. 2018. Data and systems for medication-related text classification and concept normalization from twitter: insights from the social media mining for health (smm4h)-2017 shared task. *Journal of the American Medical Informatics Association*, 25(10):1274–1283.

Kalyan Katikapalli Subramanyam and S Sangeetha. 2020. Deep contextualized medical concept normalization in social media text. *Procedia Computer Science*, 171:1353 – 1362. Third International Conference on Computing and Network Communications (CoCoNet'19).

Yoshimasa Tsuruoka, John McNaught, Jun'i; chi Tsujii, and Sophia Ananiadou. 2007. Learning string similarity measures for gene/protein name dictionary look-up using logistic regression. *Bioinformatics*, 23(20):2768–2774.

Elena Tutubalina, Zulfat Miftahutdinov, Sergey Nikolenko, and Valentin Malykh. 2018. Medical concept normalization in social media posts with recurrent neural networks. *Journal of biomedical informatics*, 84:93–102.

Guoyin Wang, Chunyuan Li, Wenlin Wang, Yizhe Zhang, Dinghan Shen, Xinyuan Zhang, Ricardo Henao, and Lawrence Carin. 2018. Joint embedding of words and labels for text classification. In *Proceedings of the 56th Annual Meeting of the Association for Computational Linguistics (Volume 1: Long Papers)*, pages 2321–2331.

Thomas Wolf, Lysandre Debut, Victor Sanh, Julien Chaumond, Clement Delangue, Anthony Moi, Pierric Cistac, Tim Rault, Rémi Louf, Morgan Funtowicz, et al. 2019. Huggingface's transformers: State-of-the-art natural language processing. *ArXiv*, pages arXiv–1910.

Maryam Zolnoori, Kin Wah Fung, Timothy B Patrick, Paul Fontelo, Hadi Kharrazi, Anthony Faiola, Yi Shuan Shirley Wu, Christina E Eldredge, Jake Luo, Mike Conway, et al. 2019. A systematic approach for developing a corpus of patient reported adverse drug events: a case study for ssri and snri medications. *Journal of biomedical informatics*, 90:103091.

Not a cute stroke: Analysis of Rule- and Neural Network-Based Information Extraction Systems for Brain Radiology Reports

Andreas Grivas[†]   **Beatrice Alex**[‡†]   **Claire Grover**[†]   **Richard Tobin**[†]   **William Whiteley** [§]

[†]School of Informatics, [‡]Edinburgh Futures Institute, [§]Centre for Clinical Brain Sciences, [§]Usher Institute
University of Edinburgh, United Kingdom
[§]Nuffield Department of Population Health
University of Oxford, United Kingdom
`{agrivas|balex|grover|richard}@inf.ed.ac.uk`

Abstract

We present an in-depth comparison of three clinical information extraction (IE) systems designed to perform entity recognition and negation detection on brain imaging reports: EdIE-R, a bespoke rule-based system, and two neural network models, EdIE-BiLSTM and EdIE-BERT, both multi-task learning models with a BiLSTM and BERT encoder respectively. We compare our models both on an in-sample and an out-of-sample dataset containing mentions of stroke findings and draw on our error analysis to suggest improvements for effective annotation when building clinical NLP models for a new domain. Our analysis finds that our rule-based system outperforms the neural models on both datasets and seems to generalise to the out-of-sample dataset. On the other hand, the neural models do not generalise negation to the out-of-sample dataset, despite metrics on the in-sample dataset suggesting otherwise.

1 Introduction

Information Extraction (IE) from radiology reports is of great interest to clinicians given its potential for automating large scale data linkage, targeted cohort selection, retrospective statistical analyses, and clinical decision support (Pons et al., 2016). Accurate IE from radiology reports has also received a surge of attention due to the insatiable demand of deep learning medical image classifiers for more labelled training data (Irvin et al., 2019).

While IE from radiology reports is of increasing value, the scarcity of annotated data and limited transferability of previously developed models is currently hindering progress. Despite recent breakthroughs in learning contextual representations for clinical and biomedical text from large amounts of unlabelled text (Devlin et al., 2019; Peng et al., 2019; Alsentzer et al., 2019; Lee et al., 2019), labelled data scarcity remains the bottleneck to improvements and wider adoption of deep learning

methods. Data scarcity is even more prominent in the general clinical domain with its vast quantity of possible entity labels.

Existing approaches to overcome the lack of labelled data include using a rule-based system to annotate more data (Smit et al., 2020) or propose labels in an annotation tool (Nandhakumar et al., 2017; Alex et al., 2019; Searle et al., 2019), leveraging semi-supervised learning to speed up annotation (Wood et al., 2020) and creating artificial data (Schrempf et al., 2020). It is also common for rule-based systems to be developed alongside statistical models to contrast their performance (Cornegruta et al., 2016; Gorinski et al., 2019; Sykes et al., 2020). We need to understand the shortcomings and benefits of rule-based and neural models to improve annotation decisions and system evaluation, a comparison which we explore in this paper both on in- and out-of-sample data.

The use of end-to-end learning for document labelling has been a recent trend in analysing radiology reports (Smit et al., 2020; Schrempf et al., 2020). Contextual representations of a document such as Bidirectional Encoder Representations from Transformers (BERT) (Devlin et al., 2019) are used as input to a multi-label classifier to label the report directly without first recognising named entities and negation. While such approaches make annotation simpler and faster and rely less on complex modelling decisions, they have various shortcomings. Firstly, they lack in interpretability, as it is hard to probe which parts of a document the model uses when making predictions. Some models employ an attention mechanism highlighting tokens in the input used to arrive at the decision (Mullenbach et al., 2018; Schrempf et al., 2020). However, they are opaque as to the exact sub-decisions that lead to the labels, which is unsatisfactory in the clinical domain where interpretability is of paramount importance. Secondly,

Proceedings of the 11th International Workshop on Health Text Mining and Information Analysis, pages 24–37
November 20, 2020. ©2020 Association for Computational Linguistics
https://doi.org/10.18653/v1/P17

they are not data-efficient. For example, Smit et al. (2020) predict four labels per entity type (positive, negated, uncertain and blank). To scale such approaches to more entity types a lot of annotated data is needed, the absence of which is currently a limiting factor. Lastly, a significant drawback of end-to-end approaches is that no part of the system other than the encoder is reusable in any domain that has a different non-overlapping set of output labels. For that new domain, the labelling procedure needs to be initiated from scratch, leading to a duplication of effort.

In this work, as in some previous neural approaches (Bhatia et al., 2019) and as is common in rule-based approaches (Cornegruta et al., 2016; Fu et al., 2019), we employ a bottom up approach to document labelling by factoring the problem into sub-tasks. This way document labels are interpretable as a sequence of decisions with some sub-tasks being extendable and reusable on other datasets. Our three IE systems, EdIE-R, EdIE-BiLSTM and EdIE-BERT (a rule-based and two neural models), recognise mentions of stroke, stroke sub-types and other related *findings* such as tumours and small vessel disease in text. They also identify related temporal *modifiers* (recent or old) and location *modifiers* (deep or cortical). For downstream document classification by phenotype, the systems also mark findings and modifier entities for negation (*negation detection*).

The contributions of our work are three-fold:

1. We compare our systems both on an in-sample and an out-of-sample dataset, drawing attention to generalisation issues of our neural models' negation detection on the out-of-sample dataset which are opaque when inspecting metrics on the in-sample one.

2. We draw on our error analysis to highlight ways in which using previously developed systems to suggest labels for new data can go wrong and propose using pretrained neural contextual models, such as BERT, to detect and correct inconsistencies.

3. We make our code[1], models and web interface[2] publicly available for re-use on brain imaging reports, as a way to bring the software to the data and assist research in this area.

---

[1] https://github.com/Edinburgh-LTG/edieviz
[2] http://jekyll.inf.ed.ac.uk/edieviz/

2 Related Work

Named entity recognition (NER) is a standard natural language processing (NLP) task and is commonly limited to identifying proper nouns in text (e.g., *person*, *organisation*, and *location*) (Sang and Meulder, 2003). In the clinical domain concepts of interest are usually *problems*, *tests* and *treatments*, as formulated in the clinical concept extraction i2b2 shared task (Uzuner et al., 2011). In our case, as in previous work on text mining and IE applied to radiology reports (Hassanpour and Langlotz, 2016; Cornegruta et al., 2016; Zhu et al., 2019), we use NER to refer to recognising entities that are either relevant medical *findings*, such as *ischemic stroke*, or *modifiers*, such as *acute*.

Approaches for NER in this domain, while not mutually exclusive, can broadly be categorised into the following: approaches leveraging lexicons, such as cTAKES (Savova et al., 2010) and RadLex (Langlotz, 2006); ontologies, such as MetaMap (Aronson and Lang, 2010); rule-based systems and pattern matching (Cornegruta et al., 2016); feature based machine learning such as Conditional Random Fields (CRFs) (Hassanpour and Langlotz, 2016); and more recently, deep learning (Cornegruta et al., 2016; Zhu et al., 2019).

Negation detection is commonly framed as identifying negation or speculation cues and their matching scopes in sentences (Fancellu et al., 2017). In the clinical domain, however, it is common for approaches to tackle negation assertion, namely, to verify whether each identified entity mention in the text is negated or affirmed (Bhatia et al., 2019), and in some cases, whether it is uncertain (Peng et al., 2018), conditionally present, hypothetically present or relating to some other patient (Uzuner et al., 2011).

As with NER, some of the earlier negation detection approaches were rule-based. NegEx (Chapman et al., 2001) relies on regular expressions to detect negation patterns, and has been successfully applied to discharge summaries. Hassanpour and Langlotz (2016) and Cornegruta et al. (2016) use NegEx for negation detection on extracted entities.

Context (Harkema et al., 2009) extends NegEx to capture hypothetical mentions, experiencer information and temporality, albeit with limited success on the latter. NegBIO (Peng et al., 2018), another rule-based negation and uncertainty detection system extended through dependency parsing information, has been shown to outperform NegEx.

Similarly, Cornegruta et al. (2016) demonstrated that enhancing NegEx with Stanford dependencies outperformed their bidirectional LSTM (BiLSTM) negation model. BiLSTM approaches for negation detection have been successful, with Fancellu et al. (2017) reporting state of the art results for Bio-Scope (Vincze et al., 2008) abstracts. Sergeeva et al. (2019) outperformed the latter using pre-trained transformer models.

Despite the amount of progress on negation detection for clinical texts, however, there is still ample evidence that while fitting systems on a particular dataset is straightforward, generalising negation detection across datasets is challenging (Wu et al., 2014). This is true both for out-of-domain evaluation, such as training on a dataset of medical articles with evaluation on a dataset of clinical text (Wu et al., 2014; Miller et al., 2017), as well as for out-of-sample evaluation, where the training and test datasets are from the same domain but may have differences due to different annotation style, or distribution of named entities (Sykes et al., 2020). For the in-domain but out-of-sample case, a domain fine-tuned rule based system seems to transfer well (Sykes et al., 2020). For all other cases, transfer is challenging, both for rule-based and machine-learning models (Wu et al., 2014; Miller et al., 2017; Sykes et al., 2020), with machine learning models benefiting from the addition of in-domain data to the training set. Lin et al. (2020) demonstrate that a pretrained BERT model can improve the results of domain transfer for negation detection, but the results are still lower for out-of-domain datasets than in-domain datasets if we compare to the results of earlier models in Miller et al. (2017). In our work we concur with previous findings: our neural models do not generalise negation detection across datasets, despite both datasets comprising radiology reports with stroke findings, such as acute ischemic stroke (AIS).

Document classification In our work, we formulate NER and negation detection as sub-tasks towards document classification by phenotypes and will report derived document classification results for one label (acute ischemic stroke) on a freely available data set of brain MRI radiology reports (Kim et al., 2019) with the aim of testing generalisability of our systems. Kim et al. (2019) compared different machine learning approaches on this data and found a single decision tree performed best (precision=91.1, recall=95.3, F1=93.2

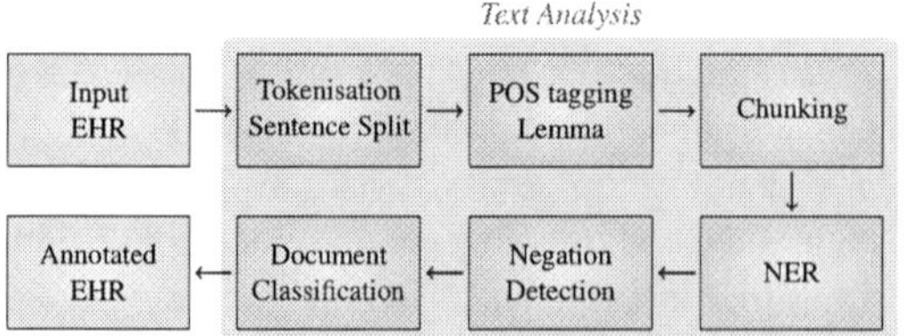

Figure 1: EdIE-R pipeline.

and acc=98%) on labelling reports as AIS or non-AIS phenotypes.

On a dataset in a similar domain, Fu et al. (2019) classify a set of 1000 radiology reports by Silent Brain Infarct. Their rule-based system based on MedTagger is superior to their word-level Convolutional Neural Network (CNN) for predicting Silent Brain Infarcts, but not for White Matter Disease.

For a broader exposition of NLP applied to radiology reports, we refer to Pons et al. (2016).

3 System Descriptions

The rule-based system, EdIE-R, and the neural systems, EdIE-BiLSTM and EdIE-BERT, all factor the document labelling task into the same three sub-tasks. Namely, extracting finding mentions, extracting modifier mentions and using negation detection to assert whether the mentions imply their presence or absence in the brain imaging report. All three systems work at a sentence level granularity.

3.1 EdIE-R

The rule-based system consists of a pipeline with four main components which are applied in sequence (see Figure 1). Two components perform linguistic analysis of the text of radiology reports, namely, NER for finding and modifier predictions and negation detection to distinguish between affirmative and negative instances. The third component computes document-level labels based on the preceding linguistic analysis. These main components are preceded by text pre-processing steps, i.e. tokenisation, part-of-speech tagging (POS) and shallow chunking.

The EdIE-R components make use of hand-crafted rules and lexicons which were created in consultation with radiology experts. The rules and lexicons are applied using the XML tools LT-XML2 (Grover and Tobin, 2006), in combination with Unix shell scripting. The NER rules are lexicon- and regular expression-dependent but the quality of the POS tagging and lemmatisation is also important. We use the C&C POS tagger (Curran and Clark, 2003) with a standard model trained on newspaper text as well as a model trained on the Genia biomedical corpus (Kim et al., 2003). After running the POS tagger with each of the models, we apply a rule-based correction stage to moderate disagreements between them. After POS tagging, we apply the *morpha* lemmatiser (Minnen et al., 2000) to analyse inflected nouns and verbs and compute their lemmas. The negation detection component relies partly on the pre-processing (i.e. recognition of negation-bearing tokens such as *no, not, n't*), and partly on the output of the chunker, which is used to constrain the scope of negative particles. The neural models we introduce in the next section rely on EdIE-R's preprocessing pipeline.

3.2 EdIE-BiLSTM

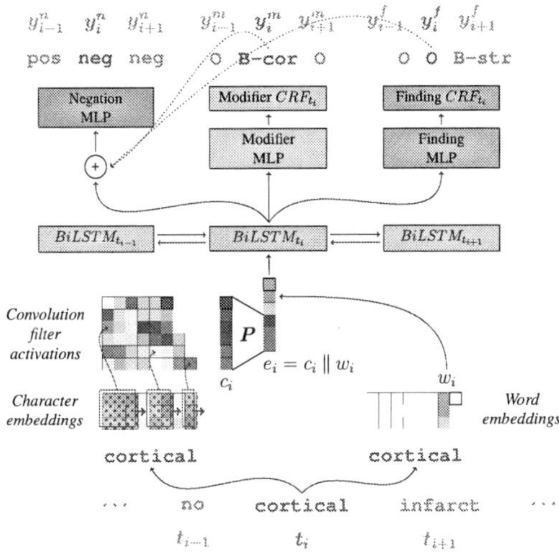

Figure 2: EdIE-BiLSTM model with multitask output for negation, finding and modifier prediction. Current input t_i and outputs y_i^{task}, such as B-cortical, are highlighted. Other timesteps appear in grey.

In order to predict the spans of findings and modifiers jointly with whether they are negated, we frame these sub-tasks as an instance of multitask learning (Caruana, 1997), similar to Bhatia et al. (2019), and train a neural network model. The network depicted in Figure 2 has three task heads, with a Conditional Random Field (CRF) output for modifiers and findings and a sigmoid binary classifier for negation. To condition negation on finding and modifier predictions, we feed the predicted findings and modifiers to the negation Multilayer Perceptron (MLP) by adding a learned embedding to the activations of the tokens that have been tagged as findings or modifiers. We do not encode entity type

to avoid biasing our negation detector towards using type:negation correlations, since as we shall see, such biases do not transfer across datasets. We decide negation for tagged entities by assigning the negation prediction made for the entity's first token. The part of the architecture mentioned so far is the same for both EdIE-BiLSTM and EdIE-BERT, the two models differ solely as to their choice of sentence encoder.

The encoder for EdIE-BiLSTM is a character CNN - word BiLSTM with randomly initialised embeddings. Given such an initialisation, it has no preconceptions about the text in the dataset and can flexibly fit the data, with the risk of overfitting. We obtain character aware token embeddings c_i by using a character level convolutional network following a modified version of the small CNN encoder model of Kim et al. (2016) (see Appendix 3.2 for details). A word-level embedding is obtained by concatenating a projected character-level token representation with a word embedding $e_i = c_i \parallel w_i$. Context-aware representations for sentence tokens are computed by propagating word-level representations through a BiLSTM network.

3.3 EdIE-BERT

EdIE-BERT only differs from EdIE-BiLSTM by replacing the BiLSTM encoder with a pretrained BERT (Devlin et al., 2019) encoder. More specifically, since we are working with radiology reports, we elected to use BlueBERT (Peng et al., 2019), which is a BERT model that is adapted for the clinical domain by being pretrained further on PubMed biomedical abstracts and clinical texts from the MIMIC-III dataset (Johnson et al., 2016). While there is a menagerie of similar models to Blue-BERT, such as BioBERT (Lee et al., 2019) and ClinicalBERT (Alsentzer et al., 2019), BlueBERT was found to outperform them when used for radiology report document classification (Smit et al., 2020). As a pretrained alternative, the BlueBERT encoder comes with preconceptions about clinical text. For example, synonyms occurring in similar contexts are likely to have similar representations and hence be assigned similar predictions by the classification layer. We shall see that this results in EdIE-BERT having increased recall but also many false positives because of flagging similar concepts that were not annotated in the data.

4 In-Sample Evaluation

In this section we describe the dataset we used to develop and fit our systems and report in-sample performance of our models on the unseen test set.

	ESS Dev	ESS Test
Reports	364	266
Sentences	3,837	2,855
Tokens	32,229	22,842
Findings	2,373	1,494
Modifiers	1,959	1,430
Total Entities	4,332	2,924

Table 1: ESS data statistics.

4.1 Edinburgh Stroke Study (ESS) Dataset

The ESS dataset (Alex et al., 2019) is comprised of English text reports produced by radiologists which describe findings in imaging reports. The reports are predominantly computerised tomography (CT) brain imaging reports with fewer magnetic resonance imaging (MRI) scans collected for a regional stroke study.

An example of a radiology report can be seen in Figure 7 in the Appendix. The language of radiology reports is usually short and descriptive as it is limited to descriptions of the image. Negation is usually overt (Sykes et al., 2020), e.g. *"no visible infarct"* with occasional hedging, e.g. *"there may be early signs of deterioration"*. There is some variation in radiologist styles, some use note style, others use full sentences.

Manual annotation of the reports was accomplished in tranches by two experts, a neurologist and a radiologist, correcting output of an early version of EdIE-R. The data was split into development (dev) and test data (see Table 1). Annotations include different entity types (12 finding types and 4 modifier types), relations between corresponding modifier and finding entities, negation of entities, and 24 document level labels (phenotypes). We note that negation labels are binary and are only assigned to findings or modifier entities. Annotators were instructed to mark any mention of findings and modifiers not clearly indicated to be present as negated. This paper focuses only on the entity and negation annotation. The entire ESS test set was doubly annotated to allow us to calculate inter-annotator agreement (IAA) using precision, recall and F1. IAA F1 is 96.15 for findings and 97.83 for modifiers. The combined NER and negation IAA F1 is 96.11.

The version of EdIE-R presented here was fur-

System	Task	P	R	F1
EdIE-R	Mod	**97.23**	95.73	**96.48**
	Find	**90.67**	95.58	**93.06**
	Neg	**92.46**	94.32	**93.38**
EdIE-BiLSTM	Mod	94.99	95.38	95.18
	Find	90.54	94.85	92.64
	Neg	91.04	93.43	92.22
EdIE-BERT	Mod	94.66	**96.71**	95.68
	Find	86.06	95.05	90.33
	Neg	88.94	**94.63**	91.70

Table 2: Results for predicting finding (Find) and modifier (Mod) entities as well as their negation (Neg) in the ESS test set. Best system per task in bold.

ther optimised on ESS dev. EdIE-BiLSTM and EdIE-BERT were trained using 285/364 ($\approx$80%) of ESS dev reports, with validation and hyperparameter tuning performed on the remaining 79 ($\approx$20%) reports. We report results on the unseen ESS test set which was not used for system development and hyperparameter tuning.

We used CoNLL scoring which considers a system annotation as true positive only if both the entity span and the label are correct as represented in IOB encoding (Sang and Meulder, 2003). Negation detection F1-score is computed for the predicted findings and modifiers, and hence includes error propagation from those tasks. F1-scores are computed using precision (P) and recall (R) based on the number of true positives (TP), false positives (FP) and false negatives (FN).

4.2 Results

The performance of all models is high, but EdIE-R outperforms both neural models in precision and F1-score on all sub-tasks (Table 2). EdIE-BiLSTM outperforms EdIE-BERT in F-score at detecting findings. We hypothesise this is because randomly initialised embeddings have little prior bias and can fit any potential annotation inconsistencies unhindered. Lastly, EdIE-BERT has lower precision but high recall, which suggests that the model overzealously flags plausible spans as findings.

Our results so far seem to suggest that EdIE-BERT is the worst performing model overall for detecting findings. This comes as a surprise, since other models using BlueBERT have reported state of the art results on many tasks (Peng et al., 2019; Smit et al., 2020). However, as we shall see in the error analysis of Section 6, its errors are mostly false positives and span-mismatch errors. When looking at EdIE-BERT output, a large part of the errors are plausible and may be spans that were missed or have boundaries that were annotated inconsistently.

We also note that both neural models underperform EdIE-R on negation, but not by a lot. They seem to generalise sensibly to the test set based on the ≈ 92 F1 score, but as we shall see in the next section, this in-sample high score is misleading.

5 Out-of-Sample Evaluation

We now test how well each of our systems generalises to unseen radiology reports from a different source, highlighting that our neural models do not generalise negation detection to this other dataset. By out-of-sample, we mean this dataset has similar labels but comes from a different distribution than the one the systems were developed on. We emphasise that we have not trained or adapted our models to this dataset.

5.1 AIS Dataset

We evaluate all three systems on a dataset of brain MRI reports labelled for AIS, collected at Hallym University Chuncheon Sacred Heart Hospital in South Korea and made publicly available in Kim et al. (2019). The data is labelled with binary AIS labels at the report level which correspond to the presence or absence of AIS in the report.

The data contains reports for 432 patients with MRI readings of confirmed AIS. To create it, a neuroradiologist read MRI images, and the labelling of the corresponding reports as AIS or non-AIS was derived from these readings. The 2,592 non-AIS reports are from patients who underwent MRI brain imaging for a variety of reasons not related to ischaemic stroke. Kim et al.'s training set (70%) contains 303 AIS and 1,815 non-AIS reports, and their test set (30%) contains 129 AIS and 777 non-AIS reports. We note that the non-AIS reports are from MRI scans that were carried out for non-stroke related reasons, which likely makes this task much easier than in the general setting. Since the data is shared as one file containing all reports without specifying the exact split, we used the combined train and test data for our experiment (see Table 3).

When testing on the AIS data we compute precision, recall and F1 (and other metrics reported by Kim et al., 2019) but, in contrast to the ESS data, we are dealing with document label predictions. We inferred AIS and non-AIS labels based on whether there was a sentence in the report which contained both an ischaemic stroke finding and an

	AIS Train & Test Data
Reports	3,024
Sentences	22,280
Tokens	168,718
AIS labelled reports	432
non-AIS labelled reports	2,592

Table 3: AIS data statistics. The sentence and token figures are determined using the EdIE-R tokenisation and sentence detection.

acute modifier (AIS), or not (non-AIS). Our temporal modifier *time recent* overlaps well with the use of *"acute"* in the AIS data, with the exception of the term *"sub-acute"*. For the purpose of inferring AIS labels, we therefore defined the acute modifier accordingly by excluding sub-acute mentions.

System	SP	NPV	P	R	F1	Acc (%)
EdIE-R	96.64	**99.56**	82.55	**97.45**	**89.38**	**96.70**
EdIE-BiLSTM	**97.92**	93.65	**82.80**	60.19	69.71	92.53
EdIE-BERT	97.18	94.56	79.72	66.44	72.47	92.79
Kim et. al 2019 on 30% of the data	98.50	99.20	91.10	95.30	93.20	98.00

Table 4: Results for classifying reports as AIS. We report the same metrics as Kim et al. 2019 but on all of the AIS data: Specificity (SP), Negative Predictive Value (NPV), Precision (P≡Positive Predictive Value), Recall (R≡Sensitivity), F1-score and Accuracy (Acc).

5.2 Results

Table 4 shows that EdIE-R achieves an F1-score of 89.38. The results are lower than the best results reported in Kim et al. (2019), but this is partly to be expected since we do not adapt any of our systems to the AIS dataset, apart from formulating the document level rules. Interestingly, EdIE-R's recall was two points higher but its precision was considerably lower. A neurologist examined some of the false positives which contributed to EdIE-R's lower precision and reported that they did, in fact, indicate acute ischaemic stroke. It is possible that in these cases AIS was not the primary finding and that these reports were therefore not labelled as AIS. Given that our systems are configured to recognise all findings in a report at the entity level, it is not surprising to find a difference in predictions as compared to a binary document labelling system, but we consider the EdIE-R results to be an effective validation of our approach and can show that it generalises to other similar data.

Both neural systems had much worse results than EdIE-R, mostly due to considerably lower recall, demonstrating poor generalisation. On inspection of their predictions, we found that this was overwhelmingly due to errors in negation detection. When removing negation, the results were very similar to those of EdIE-R. We found that one reason for the discrepancy is that the distribution of negation over findings in the ESS dataset compared to AIS is very different, with acute ischemic stroke being negated much more often in the ESS dataset compared to the AIS dataset. Despite not providing finding and modifier information to the negation detection head explicitly, the neural models seem to be using superficial features such as the distribution of negation for *acute* and *ischemic stroke* rather than relying on other features, such as overt negation cues, that would generalise. In this respect, our findings are similar to Fancellu et al. (2017), who demonstrated that neural network models were using punctuation as a cue for negation scope detection and failing to generalise beyond that.

6 Error Analysis

In this section we provide a fine-grained breakdown of the types of errors made by EdIE-R, EdIE-BiLSTM and EdIE-BERT, arguing that not all error types are equally detrimental to the downstream task of document labelling. Next, we investigate the variability in error types between our systems by exploiting BlueBERT's context-aware embeddings to group together training and evaluation examples that are similar. We then compare their labels to identify annotation artefacts that influence system errors. Lastly, we investigate how our systems handle spelling errors.

6.1 Breakdown of Error Types

As alluded to in Section 4.2, CoNLL F1 score harshly penalises wrong entity boundaries by reducing both precision and recall simultaneously (Finkel et al., 2005). For a deeper understanding of the situation, we dissect the errors (FP and FN counts) on the ESS test set into the following types (Manning, 2006):

False Positive (FP): predicted spurious entity
False Negative (FN): missed gold entity
Label Error (LE): correct span, wrong label
Boundary Error (BE): span overlap, correct label
Label & Boundary Error (LBE): span overlap + LE

We note that when relying on finding and modifier predictions for document classification by phenotype, some errors are worse than others. We

System	Task	FP	FN	LE	BE	LBE
EdIE-R	Mod	33	50	0	1	5
	Find	100	20	1	37	7
EdIE-BiLSTM	Mod	45	37	3	13	5
	Find	96	27	10	34	5
EdIE-BERT	Mod	50	17	1	16	5
	Find	164	13	11	45	4

Table 5: Number of error types made by each system for findings and modifiers in ESS test set.

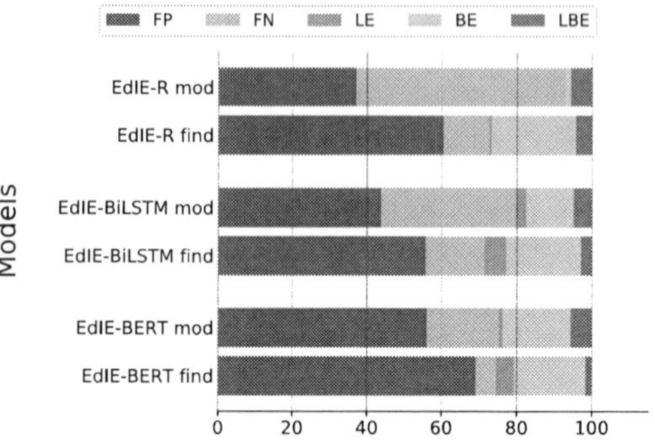

Figure 3: Proportion of error types made by each system for findings and modifiers in ESS test set.

disfavour FPs, LEs and LBEs as they are likely to deteriorate document classification by hallucinating phenotypes. We also dislike FNs, but less so, since usually radiology reports have some degree of redundancy. Lastly, we argue the BEs are mostly benign, since for document classification the span of an entity should not affect label allocation.

Table 5 and Figure 3 show that EdIE-BiLSTM and EdIE-BERT make a larger proportion of LEs, which we found to be mostly due to ambiguity in annotation between *haemorrhagic stroke* and *stroke*. There was only one BE by EdIE-R, in contrast to more than ten by EdIE-BiLSTM (13) and EdIE-BERT (16), but on recognising findings, all three systems make more than 18% BEs. The large percentage of BEs, especially on modifiers, suggest inconsistencies in span selection during annotation. Such span inconsistencies unfairly lower the score of models when evaluating by NER F1. We conclude that care is needed when relying on a subtask metric that may not correlate with the document labelling goal as well as initially expected.

A striking difference is that EdIE-BERT has a larger proportion of FPs than the other systems, with the remaining errors being mostly BEs. This highlights that the model flags multiple spurious spans that are not annotated in the data, which as we shall see in the next section, is mostly due to

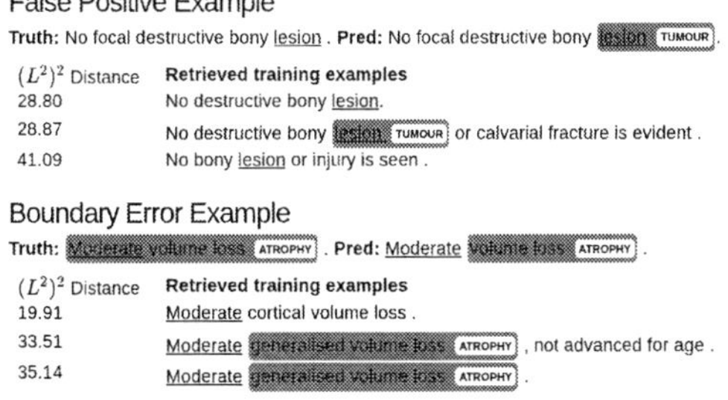

Figure 4: A EdIE-BERT *False Positive* and *Boundary Error* from the ESS test set. **Truth** and **Pred** are the gold annotations and EdIE-BERT predictions respectively. The sentences below are the three most similar training examples ordered by decreasing similarity.

inconsistencies in annotation than to model errors.

Lastly, EdIE-BiLSTM's larger proportion of FNs can be partly attributed to abbreviations, since EdIE-BiLSTM misses some not present in the training set, such as *CADASIL* (Cerebral Autosomal Dominant Arteriopathy with Subcortical Infarcts and Leukoencephalopathy) and *PICH* (Primary Intracerebral Haemorrhage). On the other hand, interestingly, EdIE-BERT tags some abbreviations that were unseen during training, such as *METS*, which can be short for *metastatic tumour*.

6.2 Nearest Neighbour Annotations

In this section, we exploit a pretrained[3] BlueBERT model's context-aware embeddings to group together sentence examples from all ESS data that are similar. We do so to gain insight into any potential annotation artefacts by contrasting the annotations of similar examples.

We follow Khandelwal et al. (2020): see equations (1) and (2) in their paper for technical details. We create a datastore with key value pairs, where the keys are BlueBERT embeddings of each token in the ESS training set and values are the token's labels. We then conduct an error analysis. For each token EdIE-BERT mislabelled during evaluation, we find the k nearest neighbour tokens from the training set and visualise their labels.

In Figure 4 we plot two examples of EdIE-BERT errors on *findings*, a FP and a BE. Above on the left is the gold annotation with the prediction on the right and the error underlined. Below are the three most similar training examples as ordered by decreasing similarity using BlueBERT[4]. In the

---

[3]Not finetuned on radiology data as part of EdIE-BERT.

[4]The nearest neighbour search is among tokens such as *lesion*, but we visualise the whole sentence for context.

FP example, we notice that *lesion* is tagged in one example as *tumour* and in others as *O*, despite the examples being very similar. In the BE example, *Moderate* is not predicted to be part of the *atrophy* finding. However, it is also not annotated as such in all similar training examples below, thus highlighting how some errors can be explained by identifying inconsistencies in annotations.

For such cases where the training set contains many alternative possible labellings of tokens in particular contexts, we propose visualising the uncertainty by plotting the entropy of the kNN distribution along the sequence together with the subset of labels deemed plausible from the retrieved training examples. Figure 5 demonstrates how the boundaries of the *small vessel disease* finding are uncertain in the training set, with some instances including *periventricular* as part of the entity, and others tagging *white* as *O* in similar contexts.

To conclude, BlueBERT's pretrained preconceptions about which contexts are similar makes it harder for the model to fit examples that are annotated inconsistently with respect to spans or labels. We believe it therefore to be an effective model for fine-grained error analysis as well as for assisting in annotation efforts in tandem with any rule-based or other developed system when generating annotations in a new domain.

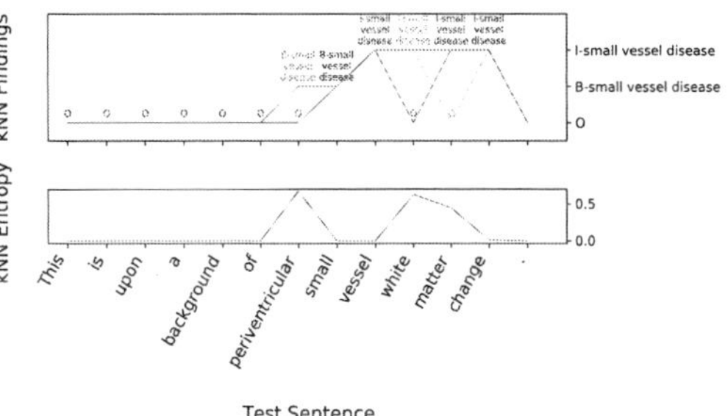

Figure 5: Distribution over *findings* p_{kNN} computed using k=10 most similar training examples to highlight uncertainty in conflicting annotations. The top subfigure demonstrates plausible labels, with solid lines linking more likely candidates. The bottom subfigure is a plot of the entropy of p_{kNN}, with higher entropy corresponding to choices that are more uncertain.

6.3 Spelling Errors

Spelling variation and spelling mistakes are not uncommon in radiology reports. For example, the ESS data contains frequent mentions of the British English spelling variant *haemorrhage* but also several mentions of *hemorrhage*, its US English version. It also includes spelling errors such as (*haemohhage*, *haemorrghage* and *heamrrhage*).

Known spelling variants can be handled with a few rules in a rule-based NLP system given a specific domain as brain imaging reports. However, terms containing spelling errors are unpredictable and hence more difficult to recognise using heuristics. Even though EdIE-R slightly outperforms the neural system overall, one main strength of EdIE-BiLSTM and EdIE-BERT is that they are robust towards spelling errors, since their model is context and subword structure-aware.

EdIE-R does not currently contain a separate spelling correction step but encodes a limited number of rules to deal with spelling errors and variations frequently observed in the data used for its development. As a result, it regards most words containing spelling errors as being out-of-vocabulary.

To examine how the neural systems dealt with actual spelling errors in radiology reports, we identified those appearing in gold findings and modifier annotations in the ESS data and found 24 unique annotations containing spelling errors (see Appendix B.1). 10 of them occur in the ESS validation and test data not used for training. Both EdIE-BiLSTM and EdIE-BERT were able to correctly recognise 6/10 and 5/10 annotations, respectively. When presented with the correctly spelled variants in the same context, they were able to identify 8/10 and 7/10 annotations accurately. While these examples are too few for a quantitative analysis of the robustness of both models towards spelling errors, it is clear that they can detect some of them accurately.

7 Summary and Conclusions

Access to annotated clinical text is a bottleneck to progress in clinical IE. While it is vital to strive for high quality gold datasets that are annotated from scratch with clear annotation guidelines, the reality of the situation is that many teams face data accessibility issues, strict time constraints and limited access to expert annotators, whose time is extremely valuable. Given finite resources, it is therefore common to leverage output from previously developed systems to speed up annotation. Through extensive error analysis, we exposed artefacts of annotations originating from experts correcting system output and recommend exploiting context-aware embedding models, such as BERT, to improve recall and ameliorate annotation inconsistencies. We are not suggesting that standards of the annotation procedure should be overlooked, but we highlight that our approach may be of value for many teams that are not in a position to label a dataset from scratch: semi-automated expert data is extremely useful under low resource settings, and therefore having a way to guide such annotation processes is valuable.

We also highlighted the pitfall of blindly trusting well-established metrics, both for ranking systems on subtasks that do not directly match the downstream task and, more importantly, in the case of generalisation, where metrics on in-sample data were misleading as to how well our neural models were capturing negation. We concur with the findings in Wu et al. (2014), negation detection is straightforward to optimise for an in-domain sample of data, but generalisation to other datasets without any adaptation is still challenging. Therefore, negation detection models should be tested across multiple datasets for generalisation.

To conclude, our rule-based system outperforms our neural network models on the limited sized in-sample dataset and generalises to an unseen dataset of radiology reports. Through a manual error analysis, we found that a large proportion of errors of our systems are due to ambiguities in annotation. Given the fairly high performance of our models, we extrapolate that we have likely distilled most of the information available in our limited labelled dataset. In future work we plan to extend our annotations to a larger dataset to further assess generalisation.

Acknowledgements

We thank Arlene Casey and the anonymous reviewers for their comprehensive feedback and Laura Perez-Beltrachini for helpful discussions on an early version of the paper. We also wish to thank Prof. Cathy Sudlow for making the ESS data available for this research and the members of the Edinburgh Clinical NLP group[5] for their support.

This research was supported by the MRC Mental Health Data Pathfinder Award (MRC - MCPC17209). Moreover, Alex and Grover have been supported by the Alan Turing Institute via Turing Fellowships (EPSRC grant EP/N510129/1). Whiteley was supported by an MRC Clinician Scientist Award (G0902303) and is supported by a Scottish Senior Clinical Fellowship (CAF/17/01).

---

[5]`https://www.ed.ac.uk/usher/`
`clinical-natural-language-processing`

References

Beatrice Alex, Claire Grover, Richard Tobin, Cathie Sudlow, Grant Mair, and William Whiteley. 2019. Text mining brain imaging reports. *Journal of Biomedical Semantics*, 10(1):23.

Emily Alsentzer, John Murphy, William Boag, Wei-Hung Weng, Di Jin, Tristan Naumann, and Matthew McDermott. 2019. Publicly available clinical BERT embeddings. In *Proceedings of the 2nd Clinical Natural Language Processing Workshop*, pages 72–78, Minneapolis, Minnesota, USA. Association for Computational Linguistics.

Alan R. Aronson and François-Michel Lang. 2010. An overview of MetaMap: historical perspective and recent advances. *Journal of the American Medical Informatics Association*, 17(3):229–236.

Parminder Bhatia, Busra Celikkaya, and Mohammed Khalilia. 2019. Joint entity extraction and assertion detection for clinical text. In *Proceedings of the 57th Annual Meeting of the Association for Computational Linguistics*, pages 954–959, Florence, Italy. Association for Computational Linguistics.

Rich Caruana. 1997. Multitask learning. *Machine Learning*, 28(1):41–75.

Wendy Chapman, Will Bridewell, Paul Hanbury, Gregory F. Cooper, and Bruce Buchanan. 2001. A simple algorithm for identifying negated findings and diseases in discharge summaries. *Journal of Biomedical Informatics*, 34:301–310.

Savelie Cornegruta, Robert Bakewell, Samuel Withey, and Giovanni Montana. 2016. Modelling radiological language with bidirectional long short-term memory networks. In *Proceedings of the Seventh International Workshop on Health Text Mining and Information Analysis*, pages 17–27, Auxtin, TX. Association for Computational Linguistics.

James Curran and Stephen Clark. 2003. Language independent NER using a maximum entropy tagger. In *Proceedings of the Seventh Conference on Natural Language Learning, CoNLL 2003, Held in cooperation with HLT-NAACL 2003, Edmonton, Canada*, pages 164–167.

Jacob Devlin, Ming-Wei Chang, Kenton Lee, and Kristina Toutanova. 2019. BERT: Pre-training of deep bidirectional transformers for language understanding. In *Proceedings of the 2019 Conference of the North American Chapter of the Association for Computational Linguistics: Human Language Technologies, Volume 1 (Long and Short Papers)*, pages 4171–4186, Minneapolis, Minnesota. Association for Computational Linguistics.

Federico Fancellu, Adam Lopez, Bonnie Webber, and Hangfeng He. 2017. Detecting negation scope is easy, except when it isn't. In *Proceedings of the 15th Conference of the European Chapter of the Association for Computational Linguistics: Volume 2, Short Papers*, pages 58–63, Valencia, Spain. Association for Computational Linguistics.

Jenny Rose Finkel, Shipra Dingare, Christopher D. Manning, Malvina Nissim, Beatrice Alex, and Claire Grover. 2005. Exploring the boundaries: gene and protein identification in biomedical text. *BMC Bioinformatics*, 6(S-1).

Sunyang Fu, Lester Y. Leung, Yanshan Wang, Anne-Olivia Raulli, David F. Kallmes, Kristin A. Kinsman, Kristoff B. Nelson, Michael S. Clark, Patrick H. Luetmer, Paul R. Kingsbury, et al. 2019. Natural language processing for the identification of silent brain infarcts from neuroimaging reports. *JMIR Medical Informatics*, 7(2):e12109.

Yarin Gal and Zoubin Ghahramani. 2016. A theoretically grounded application of dropout in recurrent neural networks. In *Advances in Neural Information Processing Systems 29: Annual Conference on Neural Information Processing Systems 2016, December 5-10, 2016, Barcelona, Spain*, pages 1019–1027.

Philip John Gorinski, Honghan Wu, Claire Grover, Richard Tobin, Conn Talbot, Heather Whalley, Cathie Sudlow, William Whiteley, and Beatrice Alex. 2019. Named entity recognition for electronic health records: A comparison of rule-based and machine learning approaches. *Computing Research Repository*, arXiv:1903.03985. Version 2.

Claire Grover and Richard Tobin. 2006. Rule-based chunking and reusability. In *Proceedings of LREC 2006*, pages 873–878.

Henk Harkema, John N. Dowling, Tyler Thornblade, and Wendy W. Chapman. 2009. Context: an algorithm for determining negation, experiencer, and temporal status from clinical reports. *Journal of biomedical informatics*, 42(5):839–851.

Saeed Hassanpour and Curtis P. Langlotz. 2016. Information extraction from multi-institutional radiology reports. *Artificial Intelligence in Medicine*, 66:29–39.

Sergey Ioffe and Christian Szegedy. 2015. Batch normalization: Accelerating deep network training by reducing internal covariate shift. volume 37 of *Proceedings of Machine Learning Research*, pages 448–456, Lille, France. PMLR.

Jeremy Irvin, Pranav Rajpurkar, Michael Ko, Yifan Yu, Silviana Ciurea-Ilcus, Chris Chute, Henrik Marklund, Behzad Haghgoo, Robyn Ball, Katie Shpanskaya, et al. 2019. Chexpert: A large chest radiograph dataset with uncertainty labels and expert comparison. In *Proceedings of the AAAI Conference on Artificial Intelligence*, volume 33, pages 590–597.

Alistair EW Johnson, Tom J Pollard, Lu Shen, H Lehman Li-wei, Mengling Feng, Mohammad Ghassemi, Benjamin Moody, Peter Szolovits, Leo Anthony Celi, and Roger G Mark. 2016.

MIMIC-III, a freely accessible critical care database. *Scientific data*, 3:160035.

Urvashi Khandelwal, Omer Levy, Dan Jurafsky, Luke Zettlemoyer, and Mike Lewis. 2020. Generalization through memorization: Nearest neighbor language models. In *International Conference on Learning Representations (ICLR)*.

Chulho Kim, Vivienne Zhu, Jihad Obeid, and Leslie Lenert. 2019. Natural language processing and machine learning algorithm to identify brain MRI reports with acute ischemic stroke. *PLOS ONE*, 14(2):1–13.

J.-D. Kim, T. Ohta, Y. Tateisi, and J. Tsujii. 2003. GE-NIA corpus—a semantically annotated corpus for bio-textmining. *Bioinformatics*, 19(1):180–182.

Yoon Kim, Yacine Jernite, David A. Sontag, and Alexander M. Rush. 2016. Character-aware neural language models. In *Proceedings of the Thirtieth AAAI Conference on Artificial Intelligence, February 12-17, 2016, Phoenix, Arizona, USA*, pages 2741–2749. AAAI Press.

Curtis P. Langlotz. 2006. Radlex: a new method for indexing online educational materials. *Radiographics*, 26(6):1595–1597.

Jinhyuk Lee, Wonjin Yoon, Sungdong Kim, Donghyeon Kim, Sunkyu Kim, Chan Ho So, and Jaewoo Kang. 2019. BioBERT: a pretrained biomedical language representation model for biomedical text mining. *Bioinformatics*, 36(4):1234–1240.

Chen Lin, Steven Bethard, Dmitriy Dligach, Farig Sadeque, Guergana Savova, and Timothy A Miller. 2020. Does BERT need domain adaptation for clinical negation detection? *Journal of the American Medical Informatics Association*, 27(4):584–591.

Christopher Manning. 2006. Doing Named Entity Recognition? Don't optimize for F1. https://nlpers.blogspot.com/2006/08/doing-named-entity-recognition-dont.html[Accessed: 29/01/2020].

Timothy Miller, Steven Bethard, Hadi Amiri, and Guergana Savova. 2017. Unsupervised domain adaptation for clinical negation detection. In *BioNLP 2017*, pages 165–170, Vancouver, Canada,. Association for Computational Linguistics.

Guido Minnen, John Carroll, and Darren Pearce. 2000. Robust, applied morphological generation. In *Proceedings of INLG 2000*, pages 201–208.

James Mullenbach, Sarah Wiegreffe, Jon Duke, Jimeng Sun, and Jacob Eisenstein. 2018. Explainable prediction of medical codes from clinical text. In *Proceedings of the 2018 Conference of the North American Chapter of the Association for Computational Linguistics: Human Language Technologies, Volume 1 (Long Papers)*, pages 1101–1111, New Orleans, Louisiana. Association for Computational Linguistics.

Nidhin Nandhakumar, Ehsan Sherkat, Evangelos E. Milios, Hong Gu, and Michael Butler. 2017. Clinically significant information extraction from radiology reports. In *Proceedings of the 2017 ACM Symposium on Document Engineering*, DocEng '17, page 153–162, New York, NY, USA. Association for Computing Machinery.

Adam Paszke, Sam Gross, Francisco Massa, Adam Lerer, James Bradbury, Gregory Chanan, Trevor Killeen, Zeming Lin, Natalia Gimelshein, Luca Antiga, Alban Desmaison, Andreas Köpf, Edward Yang, Zachary DeVito, Martin Raison, Alykhan Tejani, Sasank Chilamkurthy, Benoit Steiner, Lu Fang, Junjie Bai, and Soumith Chintala. 2019. Pytorch: An imperative style, high-performance deep learning library. In *Advances in Neural Information Processing Systems 32: Annual Conference on Neural Information Processing Systems 2019, NeurIPS 2019, 8-14 December 2019, Vancouver, BC, Canada*, pages 8024–8035.

Yifan Peng, Xiaosong Wang, Le Lu, Mohammadhadi Bagheri, Ronald Summers, and Zhiyong Lu. 2018. Negbio: a high-performance tool for negation and uncertainty detection in radiology reports. *AMIA Joint Summits on Translational Science proceedings. AMIA Joint Summits on Translational Science*, 2017:188—196.

Yifan Peng, Shankai Yan, and Zhiyong Lu. 2019. Transfer learning in biomedical natural language processing: An evaluation of BERT and ELMo on ten benchmarking datasets. In *Proceedings of the 18th BioNLP Workshop and Shared Task*, pages 58–65, Florence, Italy. Association for Computational Linguistics.

Ewoud Pons, Loes M. M. Braun, M. G. Myriam Hunink, and Jan A. Kors. 2016. Natural Language Processing in Radiology: A Systematic Review. *Radiology*, 279(2):329–343.

Erik F. Tjong Kim Sang and Fien De Meulder. 2003. Introduction to the CoNLL-2003 shared task: Language-independent named entity recognition. In *Proceedings of the Seventh Conference on Natural Language Learning, CoNLL 2003, Held in cooperation with HLT-NAACL 2003, Edmonton, Canada*, pages 142–147.

Guergana K. Savova, James J. Masanz, Philip V. Ogren, Jiaping Zheng, Sunghwan Sohn, Karin C. Kipper-Schuler, and Christopher G. Chute. 2010. Mayo clinical Text Analysis and Knowledge Extraction System (cTAKES): architecture, component evaluation and applications. *Journal of the American Medical Informatics Association*, 17(5):507–513.

Patrick Schrempf, Hannah Watson, Shadia Mikhael, Maciej Pajak, Matúš Falis, Aneta Lisowska, Keith W. Muir, David Harris-Birtill, and Alison Q.

O'Neil. 2020. Paying per-label attention for multi-label extraction from radiology reports. In *Interpretable and Annotation-Efficient Learning for Medical Image Computing*, pages 277–289, Cham. Springer International Publishing.

Thomas Searle, Zeljko Kraljevic, Rebecca Bendayan, Daniel Bean, and Richard Dobson. 2019. MedCAT-Trainer: A biomedical free text annotation interface with active learning and research use case specific customisation. In *Proceedings of the 2019 Conference on Empirical Methods in Natural Language Processing and the 9th International Joint Conference on Natural Language Processing (EMNLP-IJCNLP): System Demonstrations*, pages 139–144, Hong Kong, China. Association for Computational Linguistics.

Elena Sergeeva, Henghui Zhu, Amir Tahmasebi, and Peter Szolovits. 2019. Neural token representations and negation and speculation scope detection in biomedical and general domain text. In *Proceedings of the Tenth International Workshop on Health Text Mining and Information Analysis (LOUHI 2019)*, pages 178–187, Hong Kong. Association for Computational Linguistics.

Akshay Smit, Saahil Jain, Pranav Rajpurkar, Anuj Pareek, Andrew Y. Ng, and Matthew P. Lungren. 2020. Chexbert: Combining automatic labelers and expert annotations for accurate radiology report labeling using bert. *Computing Research Repository*, arXiv:2004.09167. Version 2.

Dominic Sykes, Andreas Grivas, Claire Grover, Richard Tobin, Cathie Sudlow, William Whiteley, Andrew MacIntosh, Heather Whalley, and Beatrice Alex. 2020. Comparison of Rule-Based and Neural Network Models for Negation Detection in Radiology Reports. *Journal of Natural Language Engineering*. Accepted for publication.

Özlem Uzuner, Brett R. South, Shuying Shen, and Scott L. DuVall. 2011. 2010 i2b2/VA challenge on concepts, assertions, and relations in clinical text. *Journal of the American Medical Informatics Association*, 18(5):552–556.

Veronika Vincze, György Szarvas, Richárd Farkas, György Móra, and János Csirik. 2008. The BioScope corpus: biomedical texts annotated for uncertainty, negation and their scopes. *BMC bioinformatics*, 9(11):1–9.

Thomas Wolf, Lysandre Debut, Victor Sanh, Julien Chaumond, Clement Delangue, Anthony Moi, Pierric Cistac, Tim Rault, Rémi Louf, Morgan Funtowicz, Joe Davison, Sam Shleifer, Patrick von Platen, Clara Ma, Yacine Jernite, Julien Plu, Canwen Xu, Teven Le Scao, Sylvain Gugger, Mariama Drame, Quentin Lhoest, and Alexander M. Rush. 2020. Huggingface's transformers: State-of-the-art natural language processing. *Computing Research Repository*, arXiv:1910.03771. Version 4.

David A. Wood, Jeremy Lynch, Sina Kafiabadi, Emily Guilhem, Aisha Al Busaidi, Antanas Montvila, Thomas Varsavsky, Juveria Siddiqui, Naveen Gadapa, Matthew Townend, Martin Kiik, Keena Patel, Gareth Barker, Sebastian Ourselin, James H. Cole, and Thomas C. Booth. 2020. Automated labelling using an attention model for radiology reports of MRI scans (ALARM). volume 121 of *Proceedings of Machine Learning Research*, pages 811–826, Montreal, QC, Canada. PMLR.

Stephen Wu, Timothy Miller, James Masanz, Matt Coarr, Scott Halgrim, David Carrell, and Cheryl Clark. 2014. Negation's not solved: Generalizability versus optimizability in clinical natural language processing. *PLOS ONE*, 9(11):1–11.

Henghui Zhu, Ioannis Ch. Paschalidis, Christopher Hall, and Amir Tahmasebi. 2019. Context-driven concept annotation in radiology reports: Anatomical phrase labeling. *AMIA Summits on Translational Science Proceedings*, 2019:232.

A Supplemental Material

A.1 Training and hyperparameter details

A.1.1 EdIE-BiLSTM

As mentioned in Section 3.2, we employ a character level convolutional network following a modified version of the small CNN encoder model of Kim et al. (2016), details of which can be seen tabulated in Table 6. We replace tanh non-linearities with ReLUs. We also remove the highway layer since we did not observe any improvements when using it. To speed up training, we apply padding-aware batch normalisation (Ioffe and Szegedy, 2015) to the convolution activations before the ReLU non-linearity.[6] We project the character aware token embedding c_i to a vector of dimensionality 128 using an affine layer P. Word embeddings are also of dimensionality 128. Both word and character embeddings are randomly initialised.

Following Gal and Ghahramani (2016), we randomly drop out word types with 0.5 probability for words. We also follow this approach for characters, but with a lower dropout rate of 0.1.

Optimisation-wise, we trained our model using stochastic gradient descent with a batch size of 16 sentences padded to maximum length, a learning rate of 1 and a linear warmup of the learning rate over the first 200 parameter updates = 1 checkpoint. Before performing backpropagation, we clip the norm of the global gradient of the parameters to 5. We stop training when entity prediction does not improve on the validation set for 10 consecutive checkpoints. Our model is implemented using PyTorch (Paszke et al., 2019).

A.1.2 EdIE-BERT

For the BlueBERT encoder we use the uncased base model trained on PubMed and MIMIC-III. We train the model using the Adam optimiser with a learning rate of $5 \cdot 10^{-5}$ and a batch size of 16. We follow the original BERT paper and train using a warmup linear schedule, increasing the learning rate linearly for the first 400 training steps (10% of training steps) until it reaches the maximum value ($5 \cdot 10^{-5}$) and then decreasing it for the remaining 90% of training steps. A step is a parameter update, namely a forward and backward propagation of a batch. 200 parameter updates roughly correspond to 15 epochs on our ESS training set. We chose the

aforementioned hyperparameter values by conducting a search over learning rate $\{5 \cdot 10^{-5}, 2 \cdot 10^{-5}\}$, batch size $\{16, 32\}$ and number of warmup steps $\{200, 400\}$ on the development set. We use the Huggingface (Wolf et al., 2020) implementation for BERT.

Hyperparameter	Value
Char embedding dim	15
Char CNN filter widths	[1, 2, 3, 4, 5, 6]
Char CNN number filters	[25, 50, 75, 100, 125, 150]
Char project dim	128
Word embedding dim	128
BiLSTM dim	512
MLP hidden dim	512
MLP dropout	0.25
Word type dropout	0.5
Char type dropout	0.1
Non-linearity	ReLU
Optimiser	SGD
Learning rate	1
Batch size	16
Gradient clipping global norm	5

Table 6: EdIE-BiLSTM hyperparameter choice.

B Spelling Errors

B.1 List of ESS Data Annotations with Spelling Errors

basal galnglia, basal ganglia, centrum semiovale, Esatblished, exta-axial collections, extra-axia collection, extraxial collection, haemorraghic transformation, infarcion, Low attenuation of perventricular white matter, microvacular ischaemia, mircovascular ischaemia, parietooccpital, perfusion defect, periventicular low attenuation, periventricualr white matter hypoattenuation, posterior cerberal artery, resticted diffusion, thebasal ganglia, craniopharyngoma, lacumar, brainstsem, occiptal and *subdural haemohhage*

B.2 Synthetic Spelling Error Analysis

The example in Figure 6 shows EdIE-Viz output of EdIE-BiLSTM for a synthetic report with a number of deliberately inserted spelling errors.[7]

The report contains misspellings due to character and whitespace insertions (*vesssel, heamorrrhage, a_cute*), character deletions (*hypoattenuation, infaret, atrophy, disease, infarcts, stroke*) or character substitutions (e→a: *pariatal*, ae→ea: *heamorrrhage*). EdIE-BiLSTM is able to recognise most of the misspelled entities, with the exception of *atrophy*. As expected, EdIE-R was only able

---

[6] We adapt the mean and variance computation of each batch to only consider tokens that do not consist of padding.

[7] EdIE-Viz is a web-based interface to our IE models (see Appendix C).

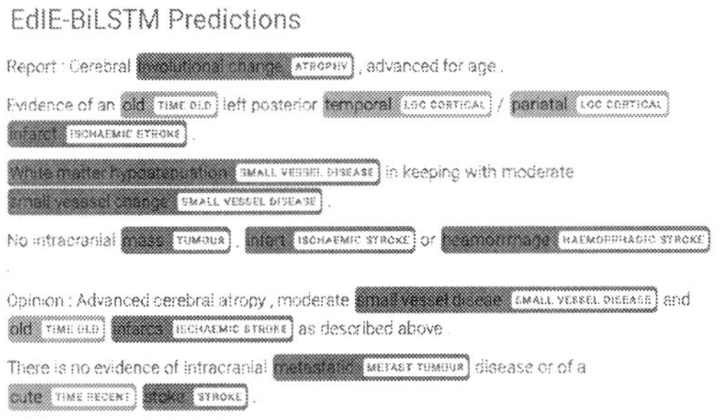

Figure 6: EdIE-BiLSTM output for a synthetic brain imaging report containing a series of spelling errors.

to tag the term *stroke* based on one of its rules allowing for that error to occur. In the case of the white-space insertion splitting *acute* into two valid English words *a cute*, EdIE-BiLSTM interestingly tags the word *cute* correctly as the temporal modifier *recent*, even though the span is wrong. Such a neural system may therefore wrongly tag a word similar in spelling but different in meaning to a medical term it is trained to extract. EdIE-BERT is able to recognise most of the misspelled findings and modifiers in this report and only differs in three cases to EdIE-BiLSTM. It is able to identify *atropy* as atrophy, does not recognise *White matter hypoatenuation* as small vessel disease and does not mark up *cute* as a modifier, presumably because during pretraining it has picked up that *cute* is a word that occurs in a different context.

C   EdIE-Viz: Interactive web demo

Our interactive web demo provides a user interface to all three systems.

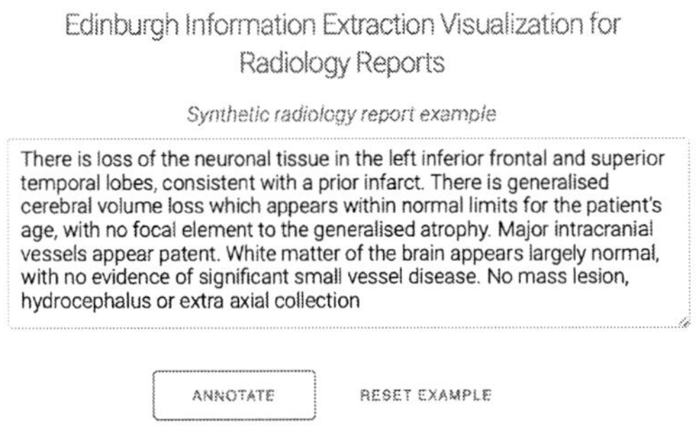

Figure 7: Home screen.

Figure 7 shows the home screen with a preloaded synthetic example of a brain imaging report. By clicking on the "Annotate" button, the demo displays[8] predicted findings (spans highlighted in

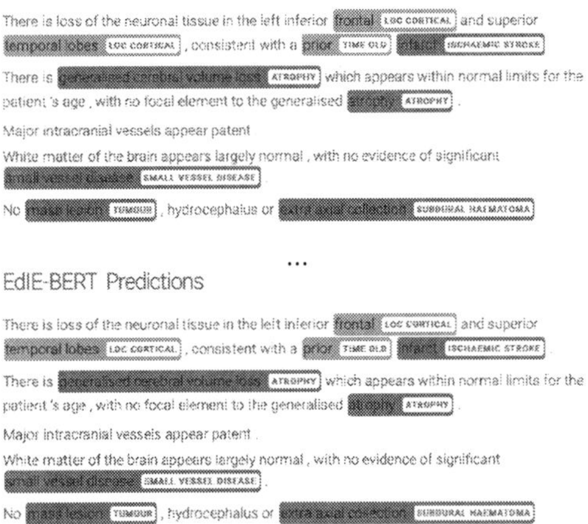

Figure 8: Predicted findings, modifiers and negation for EdIE-R and EdIE-BERT. EdIE-BiLSTM output is omitted; it is identical to that of EdIE-R for this example.

purple and types displayed behind each span in all-caps), modifiers (highlighted in orange and types in all-caps) and negation (red types for negated annotations and green types for non-negated annotations) (see Figure 8). In this example, EdIE-BERT misses the negation of *small vessel disease*.

We differentiate between findings and modifiers as they are notionally different (each modifier can be mapped to a finding) and because some tokens are tagged as both. For example, the abbreviation *POCI* (posterior circulation infarct) is tagged as *ischaemic stroke* and *cortical*.

The current use cases of this interface are the research team's own error analysis and system development, visual output analysis by example and system demonstrations to collaborators. However, in future it could be modified to allow bespoke processing of brain imaging reports, for example for assisting radiologists, or extended to add functionality that allows the comparison of other systems doing similar processing.

D   Availability of Data

The annotated ESS data has much potential value as a resource for developing text mining algorithms. This data will be available on application to Prof. Cathie Sudlow (email: Cathie.Sudlow AT ed.ac.uk) to bona fide researchers with a clear analysis plan.

---

[8]The visualisation follows the style of the displaCy Named Entity Visualiser. `spacy.io/usage/visualizers`

GGPONC: A Corpus of German Medical Text with Rich Metadata Based on Clinical Practice Guidelines

Florian Borchert[1,4*], Christina Lohr[2,5*], Luise Modersohn[2,5*], Thomas Langer[3],
Markus Follmann[3], Jan Philipp Sachs[1], Udo Hahn[2,5], and Matthieu-P. Schapranow[1,4]

[1]Digital Health Center, Hasso Plattner Institute, University of Potsdam, Germany
{firstname.lastname}@hpi.de
[2]Jena University Language & Information Engineering (JULIE) Lab
Friedrich Schiller University Jena, Germany
{firstname.lastname}@uni-jena.de
[3]German Guideline Program in Oncology, German Cancer Society, Berlin, Germany
{lastname}@krebsgesellschaft.de
[4]HiGHMED Consortium of the German Medical Informatics Initiative
[5]SMITH Consortium of the German Medical Informatics Initiative

Abstract

The lack of publicly accessible text corpora is a major obstacle for progress in natural language processing. For medical applications, unfortunately, all language communities other than English are low-resourced. In this work, we present GGPONC (German Guideline Program in Oncology NLP Corpus), a freely distributable German language corpus based on clinical practice guidelines for oncology. This corpus is one of the largest ever built from German medical documents. Unlike clinical documents, clinical guidelines do not contain any patient-related information and can therefore be used without data protection restrictions. Moreover, GGPONC is the first corpus for the German language covering diverse conditions in a large medical subfield and provides a variety of metadata, such as literature references and evidence levels. By applying and evaluating existing medical information extraction pipelines for German text, we are able to draw comparisons for the use of medical language to other corpora, medical and non-medical ones.

1 Introduction

The synthesis of validated experience in the form of Clinical Practice Guidelines (CPGs) serves as a basis for evidence-based decision making in clinical practice. To leverage the knowledge in CPGs for clinical decision support systems, e.g., for integration with electronic health records or automated evaluation of adherence to these guidelines, machine-readable versions of CPGs are necessary. However, CPGs today are disseminated mostly as free-text documents, with few formal elements. Thus, Natural Language Processing (NLP)

might be helpful to automatically extract information from these unstructured texts and transform them into a structured, or even machine-executable, format. As CPGs are also specific to their country of origin, they are usually formulated in the respective native language, so NLP technology has to be adapted properly.

A major reason for the progress in NLP research in the past years is the public availability of large text corpora (and attached metadata). Yet, for documents originating from a clinical context, the protection of personal information is a major requirement imposed by legal privacy regulations. Some research initiatives, e.g., I2B2 (Uzuner et al., 2011), MIMIC-III (Johnson et al., 2016), the Shared Task of Social Media for Health (Weissenbacher et al., 2019), or CLEF EHEALTH (Goeuriot et al., 2020) make de-identified clinical text document collections available under the conditions of Data Use Agreements (DUA). Besides, databases of biomedical research articles like PUBMED provide an abundant amount of examples for medical language. However, with only few exceptions, such easily accesssible text corpora are hardly available for the German (Lohr et al., 2018) and other non-English languages. Today, there is no viable solution for sharing even de-identified clinical texts in Germany. In effect, not only are large-scaled research datasets missing but also pretrained language models for German medical language, such as an equivalent of BioBERT (Lee et al., 2020).

To address *(1)* the lack of available German medical text resources for NLP research, and *(2)* the need for machine-readable CPGs, we constructed a corpus based on a set of German CPGs for oncology. The *German Guideline Program in Oncology*

---

*Authors marked by * equally share first authorship.

Proceedings of the 11th International Workshop on Health Text Mining and Information Analysis, pages 38–48
November 20, 2020. ©2020 Association for Computational Linguistics
https://doi.org/10.18653/v1/P17

(GGPO) (Follmann, 2020), operated by the Association of the Scientific Medical Societies in Germany, the German Cancer Society and the German Cancer Aid, is in a unique position to enable this research, as their guidelines are also provided via a mobile app (Seufferlein et al., 2019). Hence, this data set is already available in a semi-structured format with rich, formatted metadata, resulting in a much higher data quality than data extracted a posteriori from PDF versions of the guidelines.

An excerpt from the XML version of GGPONC is depicted in Listing 1. Other than clinical text relating to individual patients, we deal with scientific medical text that does not contain any privacy-sensitive data requiring de-identification. This way, we can provide access to GGPONC for other researchers via a DUA.[1]

```
<document>
 <section>
  <name>Risikofaktoren</name>
  <section>
   <name>Helicobacter pylori</name>
   <recommendation>
    <recommendation_creation_date
        value="2019-01-01"/>
    <recommendation_grade value="B"/>
    <!-- more metadata -->
    <text>Die H. pylori-Eradikation
        mit dem Ziel der Magenkarzinom
        -prävention sollte bei den
        folgenden Risikopersonen
        durchgeführt werden (siehe
        Tabelle unten).</text>
   </recommendation>
   <text>Das Magenkarzinom ist eine
        multifaktorielle Erkrankung,
        bei der die Infektion mit H.
        pylori den wichtigsten
        Risikofaktor darstellt. Seit
        1994 ist H. pylori durch die
        Weltgesund-heitsorganisation
        als Klasse I Karzinogen
        anerkannt und wurde 2009 als
        solches bestätigt <litref id="
        65327"/>.</text>
  </section> <!-- more sections -->
 </section>
</document> <!-- more documents -->
```

Listing 1: Text snippet from the XML version of the underlying GGPO corpus. Documents are structured into sections which can contain multiple recommendations. CPG recommendations can carry a multitude of metadata elements, as well as a concise text statement. Additional background text segments may contain more detailed information.

---

[1]For instructions on how to get access to the data, see: `https://www.leitlinienprogramm-onkologie.de/projekte/ggponc-english/`

2 Related work

Due to legal data protection measures, German-language clinical text corpora are extremely rare and existing ones almost impossible to (re)use. Typically, accessibility is restricted to research staff only within the lifetime of a project and blocked for the outside world. In Table 1, we list, to the best of our knowledge, all existing German-language clinical research text corpora which have been described in scientific publications up until now. With only few exceptions, these corpora are small and mostly limited to a specific medical discipline or clinical division. In addition to these pure clinical documents, other document types are also interesting for the NLP community, e.g., CPGs, which are available for a wide range of conditions.

CPGs as a target for automated text analytics have been much less utilized compared to other scientific publications and clinical documents. Most of that work took place in the context of formalizing CPGs as computer-interpretable guidelines (Peleg, 2013). Bouffier and Poibeau (2007) describe an approach to fill in a semi-structured *Guideline Elements Model* template by segmenting unstructured guidelines using linguistic patterns. An evaluation was run on 18 French guidelines. Serban et al. (2007) describe the extraction and instantiation of linguistic templates for guideline formalization, evaluated on a Dutch guideline for breast cancer treatment. German CPGs were the focus of Becker and Böckmann (2017) who adapted APACHE CTAKES to detect German UMLS concepts and evaluated their approach on a single German breast cancer guideline. Zadrozny et al. (2017) outline a system which identifies contradictions and disagreements in English CPGs.

Some authors have focused on extracting more task-specific information, such as activities (Kaiser et al., 2010), process structures (Wenzina and Kaiser, 2013; Zhu et al., 2013; Hematialam and Zadrozny, 2017) or negation triggers (Gindl et al., 2008). Taboada et al. (2013) apply a pipeline of open-source tools for parsing CPGs, Named Entity Recognition (NER) tagging and relation extraction in a case study with 171 sentences from CPGs. Most of the aforementioned approaches work with relatively small annotated corpora and English language, only. Recently, Fazlic et al. (2019) use LSTMs and fuzzy rules to extract *"action takers"*, *"symptoms"*, *"actions"* and *"purposes"* from CPGs, recognize recommendations and predict the grade

Table 1: Overview of existing text corpora of German clinical language. For GGPONC, we report the number of guidelines with the number of their individual text segments in brackets.

Corpus / Data	Documents	Sentences	Tokens	Available
FRAMED: clinical reports and medical textbook snippets (Wermter and Hahn, 2004)	–	6k	100k	✗
Reports from five medical domains (Fette et al., 2012)	544	–	–	✗
Radiology reports (Bretschneider et al., 2013)	174	4k	28k	✗
Transthoracic echocardiography reports (Toepfer et al., 2015)	140	–	–	✗
Operative reports (surgery) (Lohr and Herms, 2016)	450	22k	266k	✗
Discharge summaries from a dermatology department (Kreuzthaler et al., 2016)	1,696	–	–	✗
Discharge summaries and clinical notes from nephrology domain (Roller et al., 2016)	1,725	28k	158k	✗
Discharge summaries and clinical notes from nephrology domain (Cotik et al., 2016)	183	2k	13k	✗
X-ray reports (Krebs et al., 2017)	3,000	–	–	✗
3000PA: internistic and ICU discharge summaries 3000PA JENA PART (Hahn et al., 2018)	≈ 3,000 / 1,006	– / 170k	– / 1,421k	✗ / ✗
JSYNCC: case examples from medical textbooks (Lohr et al., 2018) (v1.1)	903	29k	368k	✓
Mixed-domain, -section, and -document type ASSESS CT corpus (Miñarro Giménez et al., 2019)	60 (400–600 chars each)	–	≈ 6k	✗
Discharge summaries with osteoporosis diagnosis (König et al., 2019)	1,982	–	2,001k	✗
Technical-Laymen Corpus: social media samples (Stomach-Intestines, Kidney) (Seiffe et al., 2020)	4,000	–	438k	✓
GGPONC – recommendations	25 (4,348)	7k	132k	✓
GGPONC – complete corpus	25 (8,418)	60k	1,340k	✓

of recommendation. The authors use a data set extracted from PDF versions of 45 guidelines with 1,020 recommendations.

Some larger corpora of CPGs for the English language exist already. Hussain et al. (2009) present the Yale Guideline Recommendation Corpus (YGRC), a sample of 1,275 guideline recommendations extracted from the *National Guideline Clearinghouse* (NGC). Their work revealed inconsistencies in writing style and reporting of the strength of recommendations. Using a subset of YGRC, El-Rab et al. (2017) present a rule-based approach to detect procedures and drug recommendations. Read et al. (2016) describe the CREST corpus, consisting of 4,029 recommendations from 170 guidelines annotated with their respective recommendation strength and report a total number of 8,138 types within the recommendations. Large corpora of CPGs lend themselves to mining the

state-of-the-art knowledge in a medical subfield. For instance, Leung et al. (2015) identify comorbidities by analyzing pairs of co-occurring conditions, using a corpus of 268 NGC guideline summaries. Leung and Dumontier (2016) find drug-disease relations via named entity recognition using a corpus of 377 NGC guideline summaries. The relations are compared to structured drug product labels to assess their overlap.

In summary, our work is most similar to the CREST corpus (Read et al., 2016), in the sense that we provide a corpus based on CPGs consisting of medical text and metadata. However, while the number of recommendations in GGPONC is comparable to CREST, the amount of structured metadata and background text in our corpus is much larger (see Section 4.1). Also, our corpus contains German text, addressing a scenario where available resources are much scarcer (see Table 1). While

Table 2: Metadata elements of recommendations of GGPONC

Attribute	Description
Recommendation creation date	Date the recommendation was first introduced
Type of recommendation	Evidence-based or consensus-based statement or recommendation
Recommendation grade	A (strong recommendation)
	B (recommendation)
	0 (weak recommendation / option)
Strength of consensus	Strong Consensus
	Consensus
	Approved by majority
	No consensus
Total vote in percentage	Percentage of approval among the expert committee
Literature references	List of evidence backing up the recommendation
Expert opinion	Yes or absent
Level of evidence	According to Oxford, SIGN, or GRADE
Edit state	State (checked, new, modified) & note regarding guideline updates

Becker and Böckmann (2017) also apply NLP to German CPGs, we consider a large superset of the CPGs used in their work and provide access to our data as a preprocessed and analyzed text corpus.

3 Methods

3.1 Data Collection

In order to assemble the corpus of German CPGs, we acquired semi-structured JSON versions of the guidelines from the REST API of the Content Management System (CMS) that serves the backend for the mobile app provided by the GGPO. The data was subsequently transformed from JSON to an XML format. We preserved the document structure (chapters and sections), as well as recommendation metadata and literature references. An example of the resulting XML format can be found in Listing 1. The metadata elements are described in Table 2. Literature references are included with an ID number which can be resolved to a citation in the provided literature index file.

The guidelines distinguish between recommendations and background texts; we preserved this distinction in the corpus. In general, the recommendations tend to be concise statements related to a particular clinical question. For evidence-based recommendations, literature references and evidence levels are included. The background texts provide the reasoning behind the recommendations and a summary of the evidence underlying the recommendations, again backed by literature references.

3.2 Automated Annotation

Besides the XML version of the corpus, we created plain text versions of all recommendations of background text parts to facilitate processing by existing NLP pipelines. For preprocessing, like sentence splitting and tokenization, we used JCORE (Hahn et al., 2016) (i.e., UIMA-based) pipelines and FRAMED (Wermter and Hahn, 2004) models which were developed for German clinical text.

We also employed the JUFIT tool (v1.1) (Hellrich et al., 2015), a filter for UMLS to create a dictionary of all German words from the UMLS (Bodenreider, 2004) (version 2019AB)[2] and the Semantic Groups ANAT (Anatomical Structure), CHEM (Chemicals & Drugs), DEVI (Devices), DISO (Disorders), LIVB (Living Beings), PHYS (Physiology), and PROC (Procedures) (without advanced JUFIT rules). We have chosen only these six out of the full set of 15 UMLS Semantic Groups because we used similar categories in the named entity recognition tasks and also wanted to avoid cognitive overloading of the human annotators.

Finally, we screened TNM expressions[3] which were extracted using a rule-based approach implemented with the PYTHON library SPACY. This part was originally developed for German pathology reports in the context of the HIGHMED consortium of the Medical Informatics Initiative of Germany. TNM expressions and genes were specifically chosen for their relevance in cancer treatment.

---

[2] https://www.nlm.nih.gov/research/umls/
[3] The UICC TNM system is a classification scheme for malignant tumors, see https://www.uicc.org/resources/tnm

Table 3: Details of GGPONC. We report the number of text **Seg**ments (plain text files), **Rec**ommendations and literature **Ref**erences. The numbers of **Sent**ences, **Tokens** and **Types** refer to the pure textual content of the corpus, excluding any meta-data and headings. Annotated parts of the corpus are marked with * (see also Section 4.3).

	Guideline	Seg.	Rec.	Sent.	Tokens	Types	Ref.
1	Palliative medicine*	696	445	5,956	134,489	15,795	3,065
2	Lung cancer*	666	313	4,251	93,324	12,756	2,344
3	Breast cancer	685	362	4,127	93,128	12,660	2,824
4	Supportive therapy	823	337	4,224	90,711	12,411	2,401
5	Bladder cancer	355	225	3,872	85,299	11,347	2,521
6	Colorectal cancer*	569	290	3,176	71,416	9,644	2,580
7	Prostate cancer	307	221	3,090	67,900	9,418	2,119
8	Malignant melanoma	297	167	2,715	60,354	9,318	1,256
9	Prevention of skin cancer	288	119	2,354	55,965	9,140	952
10	Actinic keratosis and SCC of the skin*	199	74	2,590	54,073	6,861	1,278
11	Stomach cancer	246	142	2,328	50,836	8,156	1,670
12	Endometrial cancer	317	173	1,999	50,056	8,154	1,340
13	Cervical cancer*	341	115	2,168	49,422	8,164	1,127
14	Prevention of cervix cancer*	302	103	2,055	48,676	7,989	1,391
15	Renal cell cancer*	276	122	2,118	48,013	8,202	1,496
16	Testicular tumors	315	163	1,917	43,726	6,774	1,412
17	Oesophageal cancer*	172	91	1,611	35,710	6,680	1,026
18	Laryngeal cancer	189	118	1,525	35,519	6,841	681
19	Chronic lymphocytic leukemia (CLL)*	290	138	1,410	34,470	5,682	725
20	Hodgkin lymphoma*	253	167	1,489	31,876	5,245	889
21	Hepatocellular cancer (HCC)*	157	88	1,296	27,852	5,704	803
22	Malignant ovarian tumors	193	94	1,136	25,807	5,110	1,013
23	Psycho-oncology*	121	47	778	19,270	4,127	835
24	Pancreatic cancer	294	158	857	16,871	3,670	1,154
25	Oral cavity cancer*	111	76	630	15,438	3,376	1,026
	Annotated Part	**4,153**	**2,069**	**29,528**	**664,029**	**50,732**	**18,585**
	Full Corpus	**8,414**	**4,348**	**59,672**	**1,340,201**	**76,252**	**37,928**

4   Results

4.1   Corpus Characteristics

In total, 25 GPGs with 8,414 text segments were extracted from the CMS comprising the first version of GGPONC. In Table 3, we give an overview of the CPGs in terms of the number of tokens and types, as well as the number of literature references. We also report the total number of recommendations and background text segments, since they serve as the units of analysis for our automated annotation pipelines. The CPGs cover a wide range of indications and anatomical locations. They also differ significantly in their extent, e.g., there is much more text for broad topics, such as palliative medicine, or indications with many treatment options, such as lung cancer. Of the approximately 38k literature references in the corpus, around 20k are unique with roughly 9k explicit

links to PUBMED. We provide bibliographic details on these references alongside the corpus to facilitate research on the relationships between CPGs and the underlying medical evidence.

Table 4 contains the automated named entity extraction results. Their quality and interpretation in comparison to other German (clinical and non-clinical) text corpora will be discussed in the next section. The whole corpus consists of:

- a single XML file, including the document structure and all mentioned metadata,
- a file for the complete literature index,
- individual plain text versions of the text segments, sentences, and tokens,
- automatically created entity annotations and a subset of manually corrected annotations in standoff format.

Table 4: Comparison of GGPONC with 3000PA (Jena part), JSYNCC, German PUBMED abstracts of case reports and two non-clinical corpora (German Wikipedia articles of wars (WIKIWARSDE) and news articles from the KRAUTS corpus)

		GGPONC		Clinical			Non-Clinical	
		Complete	Recom.	3000PA-J	JSYNCC	PUBMED	WIKI	KRAUTS
Documents		8,418	4,348	1106	903	336	22	142
Sentences		60k	7k	171k	29k	3k	5k	1k
Tokens		1,340k	132k	1,421k	368k	43k	96k	31k
Tokens / Sentence		22.5	19.0	8.8	12.5	16.5	20.9	25.3
UMLS*	(%)	6.42	8.93	8.72	5.71	7.59	0.75	0.02
ANAT	(%)	0.45	0.48	1.78	1.11	0.79	0.04	0.09
CHEM	(%)	0.82	1.01	1.08	0.41	0.59	0.04	0.07
DEVI	(%)	0.12	0.17	0.20	0.55	0.18	0.06	0.04
DISO	(%)	1.42	2.02	2.96	1.21	2.80	0.08	0.13
LIVB	(%)	1.07	1.32	0.38	0.35	0.82	0.38	0.37
PHYS	(%)	0.37	0.43	0.76	0.60	0.50	0.12	0.10
PROC	(%)	2.18	3.50	1.56	1.49	1.90	0.01	0.12
Genes	(%)	1.28	1.41	2.21	0.87	0.97	0.94	0.55
TNM	(%)	0.19	0.37	0.07	0.07	0.04	0.003	0
Stop words	(%)	34.05	35.53	20.37	32.96	34.51	34.65	24.24

As CPGs are subject to a regular update cycle, we are able to automatically redo the data acquisition process in the future in order to provide a historical view on the guideline development.

4.2 Comparison with Other German Medical and Non-Medical Corpora

We analyze the characteristics of GGPONC by comparing the entity matches with three German medical text corpora, namely version 1.1 of the JSYNCC corpus (case examples from clinical text books) (Lohr et al., 2018), the Jena Part of the 3000PA corpus (1106 German discharge summaries) (Hahn et al., 2018),[4] and abstracts of German case reports from PUBMED. In addition, we compare the results to out-of-domain corpora, namely German WIKIPEDIA articles of wars (WIKIWARSDE) (Strötgen and Gertz, 2011) and news articles from the KRAUTS corpus (Strötgen et al., 2018). The results are depicted in Table 4.

The fraction of stop words is comparable across all medical text corpora, as is the fraction of tokens that map to UMLS concepts. As expected, the guideline recommendations contain more medical terms per token than the background text. Compared to the clinical corpora, the CPGs have more instances of the class *Living Beings*, as they often describe treatment recommendations for certain populations. Notably, the average sentence length is much greater in the clinical guidelines, and in particular in the background text, pointing at the more scientific style of writing prevalent in the guidelines as compared to clinical narratives. TNM expressions occur much more frequently in GGPONC, which can be attributed to its focus on the oncology domain.

Both out-of-domain corpora contain only small amounts of UMLS concepts (apart from the semantic class *Living Beings*), which indicates a high precision of our entity tagging approach. In Figure 1 we visualize the overlap of unique medical concepts from UMLS found in each of the corpora.

While there is a significant overlap between GGPONC and the clinical corpora, a major fraction of concepts is unique to each corpus. These results suggest that our corpus combined with other clinical text corpora can provide a more comprehensive view on the use of medical language, in general, than each of the corpora alone.

4.3 Evaluation of Annotation Results

The automatic annotations for a subset of the CPGs have been independently reviewed by human experts (4 students of medicine, all of them passed their first medical exam, supervised by a medical

---

[4]Based on the approval by the local ethics committee (4639-12/15) and the data protection officer of Jena University Hospital discharge summaries were extracted from the HIS of the Jena University Hospital and further transformed.

Table 5: Pair-wise average F1-score and standard deviation (σ) for instance and token based inter-annotator-agreement (IAA), precision and recall per entity class. Genes had to be excluded from IAA analysis due to the large number of false positives.

		IAA Instances		IAA Token		Precision	Recall
		avg. F-score	σ	avg. F-score	σ		
UMLS Anatomy	(ANAT)	.718	.122	.720	.125	.872	.571
UMLS Chemicals	(CHEM)	.839	.045	.850	.041	.917	.600
UMLS Devices	(DEVI)	.414	.366	.409	.368	.465	.209
UMLS Disorders	(DISO)	.747	.097	.773	.096	.919	.453
UMLS Living Being	(LIVB)	.848	.066	.846	.067	.985	.698
UMLS Physiology	(PHYS)	.534	.183	.576	.177	.607	.310
UMLS Procedures	(PROC)	.706	.099	.727	.100	.944	.506
TNM (rule based)		.820	.073	.749	.120	.965	.881
Overall w/o Genes		**.742**	**.094**	**.758**	**.094**	**.945**	**.528**
Genes		-	-	-	-	.022	.589

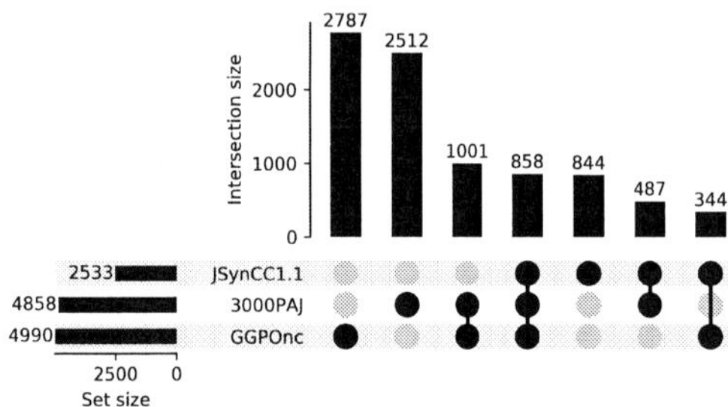

Figure 1: UPSET (Lex et al., 2014) visualization of the intersection of distinct UMLS concepts in JSYNCC 1.1, 3000PA (Jena part) and GGPONC. Each vertical bar indicates the size of one intersecting subset of UMLS terminology shared between the corpora, whereas the horizontal bars denote the total number of distinct UMLS concepts per corpus.

doctor) using the BRAT annotation tool (Stenetorp et al., 2012). Due to restricted resources for manual annotation work, we decided to evaluate on a subset of 13 (full) guidelines (see Table 3), which amounts to half of the corpus. The CPGs were chosen such that they cover a diverse range of topics and percentages of token matches, with a rather high rate of around 6–8% results per token for HCC as opposed to a lower rate of roughly 5–6% for CLL and psycho-oncology. Additional guidelines were chosen for manual annotation based on project requirements.

We calculate the inter-annotator-agreement (IAA) using the pair-wise average F-score (Hripcsak and Rothschild, 2005) of instances and tokens. An instance is a single composite annotation unit which consists of one or more tokens, e.g., *"eingeschränkte Nierenfunktion"* (limited renal function) denotes an instance with two (German) tokens, *"eingeschränkte"* and *"Nierenfunktion"*.

The agreement subset consists of 20 text segments with the largest amount of automatic annotations for each of four guidelines (HCC, CLL, Pancreatic cancer, Psycho-oncology) and 20 random-sampled text segments, resulting in 40 agreement documents with a size of approx. 19k tokens annotated by all annotators. We excluded the gene category from the IAA analysis, due to an apparently large number of false positive pre-annotations. The IAA achieved an average F-score of 0.742 on instances and 0.758 on tokens. Furthermore, we calculated micro-averaged precision and recall values for the automated annotation results, using the complete set of manually reviewed annotations as gold standard. The results are depicted in Table 5.

In another annotation study of diagnoses, symptoms and findings on the Jena part of the 3000PA corpus, average F-score values converged in the range of around 0.7–0.8 for typical clinical entities as well, e.g., anatomy or disorders in comparison to diagnoses (approx. 0.7), also for pre-annotations. The low IAA value of *Physiology* is similar to the IAA of 0.5 on the symptoms category of the named study (Lohr et al., 2020). The UMLS category *Living Beings* contains a lot of information similar to personal health information. The average IAA value of around 0.9 is similar to average values of an annotation study for the anonymization of German discharge summaries (F-score > 0.95) (Kolditz et al., 2019).

5 Discussion & Limitations of this Study

While the initial results of the information extraction pipelines we employed are promising, there is much room for improvement. The extraction of genes suffers from a large number of false positives, as there are many common German words (e.g., *"gilt"*, *"dar"*) and three-letter-acronyms (e.g., *"CLL"*, *"HCC"*) with strings identical with gene names in our large dictionary (around 562k entries). Thus, supplying an improved gene tagger which balances German lexical noise with advanced capabilities of gene taggers for English texts will be a desideratum of future research.

The German UMLS has a number of issues, which severely affect our dictionary-based entity extraction pipelines. First and foremost, its vocabulary size is extremely limited. The English UMLS contains over 6.5M entries and the Spanish one around 750k, whereas there are only around 234k entries in the German version (3.6% coverage of the English version). Recently introduced drugs are missing in the UMLS *Chemistry* category, so a more up-to-date dictionary of drug names is also needed for future work. Moreover, the surface representation of German umlauts is notoriously inconsistent in UMLS, e.g., *"ä"* is sometimes transcribed as *"ae"* or even simplified as *"a"*, as in *"eingeschraenkte Nierenfunktion"*, which results in an increasing false negative rate. All of these factors contribute to rather low recall values, as shown in Table 5.

The accuracy of dictionary matches further decreases due to inconsistent handling of compounds throughout the corpus. For instance, *"Pankreaskarzinompatienten"* (patients with pancreas carcinoma) would not be detected as an entity, whereas hyphen-connected *"Pankreaskarzinom-Patienten"* would, yielding two entities (*Disorders* and *Living Beings*), respectively. In this case, we would choose to annotate the whole compound as *Living Beings* to avoid annotation on a subword level, which could be addressed using a more finely adapted tokenization algorithm. While precision and recall of the rule-based TNM extraction approach are high on GGPONC, one has to be careful as certain TNM expressions can cause context-dependent semantic ambiguities. For instance, *"V1"* and *"V2"* are valid TNM components referring to venous invasion, but are also detected in the WIKIWARSDE corpus referring in this context to German missiles from World War II.

6 Conclusion

We presented GGPONC, one of the currently largest corpora composed of German medical texts, assembled from the CPGs in oncology and equipped with rich structure information and metadata. We applied information extraction pipelines to extract a variety of named entity classes. Despite the limitations we discussed, the information extracted so far can be of immediate use to enable semantic search functionalities in the guideline app (Seufferlein et al., 2019), precision medicine search engines (Faessler et al., 2020) or in clinical decision support systems (Schapranow et al., 2015).

Our results indicate that GGPONC shares many characteristics with existing clinical text corpora. This can facilitate the development of machine learning-based NLP algorithms for German clinical text. Beam et al. (2020) suggest that combining corpora covering different parts of medical terminology can improve the utility of trained word embeddings. In addition to the German documents discussed in this work, some of the GGPO guidelines have an additional English version, which could be used to construct parallel corpora for research in multilingual medical NLP.

Extending our work to clinical guidelines from other medical specialities besides oncology will be a straightforward way to extend the volume of the corpus, provided that the document structures can be harmonized across medical societies. However, as most CPGs are distributed as PDF documents, extraction of the plain text content from these can result in quality issues not encountered in this work.

The structured metadata of the corpus provide ample opportunities for future research. For instance, the corpus can be used as a resource for evidence-based medicine summarization, as it contains mappings from literature references to recommendation statements and evidence levels. As we plan to create future versions of the corpus based on updated guideline versions, the extracted concepts can also be used to track changes in CPGs, like the emergence of new treatments and other changes in recommended clinical practice. We envision to combine information extracted from scientific articles, such as study reports, or clinical trial registers with information from CPGs to automatically detect whether these CPGs might be outdated given changes in the underlying evidence base.

In addition to the existing annotations for a wide selection of UMLS semantic types, we can easily extend the employed pipelines with different dictionaries, e.g., derived from other subsets of the German part of the UMLS, more comprehensive official lists of drug names, or the German version of the *International Classification of Diseases*.

We make GGPONC available for researchers under the conditions of a Data Use Agreement. For instructions on how to access the corpus and the human annotated data see: `https://www.leitlinienprogramm-onkologie.de/projekte/ggponc-english/`. The code to reproduce our experiments is available at: `https://doi.org/10.5281/zenodo.4067994`.

Acknowledgments. This work was partially supported by the German Federal Ministry of Research and Education (BMBF) under grants (01ZZ1802H, 01ZZ1803G). We thank our annotators, André Scherag and Danny Ammon from the Jena University Hospital, and all colleagues from the HIGH-MED and SMITH consortia for their constant support.

References

Andrew L. Beam, Benjamin Kompa, Allen Schmaltz, Inbar Fried, Griffin Weber, Nathan Palmer, Xu Shi, Tianxi Cai, and Isaac S. Kohane. 2020. Clinical concept embeddings learned from massive sources of multimodal medical data. In *Proceedings of the Pacific Symposium on Biocomputing 2020. Big Island, Hawaii, USA, January 3-7, 2020*, pages 295–306.

Matthias Becker and Britta Böckmann. 2017. Semi-automatic mark-up and UMLS annotation of clinical guidelines. In *MedInfo 2017 — Proc. of the 16th World Congress on Medical and Health Informatics. Hangzhou, China, 21-25 Aug. 2017*, pages 294–297.

Olivier Bodenreider. 2004. The Unified Medical Language System (UMLS): integrating biomedical terminology. *Nucleic Acids Research*, 32:D267–D270.

Amanda Bouffier and Thierry Poibeau. 2007. Automatically restructuring practice guidelines using the GEM DTD. In *BioNLP 2007 — Proceedings of the Workshop on Biological, Translational, and Clinical Language Processing @ ACL 2007. Prague, Czech Republic, June 29, 2007*, pages 113–120.

Claudia Bretschneider, Sonja Zillner, and Matthias Hammon. 2013. Identifying pathological findings in German radiology reports using a syntacto-semantic parsing approach. In *BioNLP 2013 — Proceedings of the 2013 Workshop on Biomedical Natural Language Processing @ ACL 2013. Sofia, Bulgaria, August 8, 2013*, pages 27–35.

Viviana Cotik, Roland Roller, Feiyu Xu, Hans Uszkoreit, Klemens Budde, and Danilo Schmidt. 2016. Negation detection in clinical reports written in German. In *BioTxtM 2016 — Proceedings of the 5th Workshop on Building and Evaluating Resources for Biomedical Text Mining @ COLING 2016. Osaka, Japan, December 12, 2016*, pages 115–124.

Wessam Gad El-Rab, Osmar R. Zaïane, and Mohammad El-Hajj. 2017. Formalizing clinical practice guideline for clinical decision support systems. *Health Informatics Journal*, 23(2):146–156.

Erik Faessler, Michel Oleynik, and Udo Hahn. 2020. What makes a top-performing precision medicine search engine? Tracing main system features in a systematic way. In *SIGIR '20 — Proceedings of the 43rd International ACM SIGIR Conference on Research and Development in Information Retrieval. July 25-30, 2020 (Virtual Event)*, pages 459–468.

Lejla Begic Fazlic, Ahmed Hallawa, Anke Schmeink, Arne Peine, Lukas Martin, and Guido Dartmann. 2019. A novel NLP-FUZZY system prototype for information extraction from medical guidelines. In *MIPRO 2019 — Proceedings of the 42nd Intl. Convention on Information and Communication Technology, Electronics and Microelectronics. Opatija, Croatia, 20-24 May 2019*, pages 1025–1030.

Georg Fette, Maximilian Ertl, Anja Wörner, Peter Klügl, Stefan Störk, and Frank Puppe. 2012. Information extraction from unstructured electronic health records and integration into a data warehouse. In *Proceedings der 42. Jahrestagung der Gesellschaft für Informatik (GI). Braunschweig, Germany, Sept. 16-21, 2012*, pages 1237–1251.

M. Follmann. 2020. German Guideline Program in Oncology. `www.leitlinienprogramm-onkologie.de/english-language`.

José Antonio Miñarro Giménez, Ronald Cornet, Marie Christine Jaulent, Heike Dewenter, Sylvia Thun, Kirstine Rosenbeck Gøeg, Daniel Karlsson, and Stefan Schulz. 2019. Quantitative analysis of manual annotation of clinical text samples. *International Journal of Medical Informatics*, 123:37–48.

Stefan Gindl, Katharina Kaiser, and Silvia Miksch. 2008. Syntactical negation detection in clinical practice guidelines. In *MIE 2008 — Proceedings of the 21st International Congress of the European Federation for Medical Informatics. Gothenburg, Sweden, 25-28 May 2008*, pages 187–192. IOS Press.

Lorraine Goeuriot, Hanna Suominen, Liadh Kelly, Antonio Miranda-Escalada, Martin Krallinger, Zhengyang Liu, Gabriella Pasi, Gabriela Gonzalez Saez, Marco Viviani, and Chenchen Xu. 2020. Overview of the CLEF eHealth Evaluation Lab 2020. In *CLEF 2020 — Proceedings of the 11th Intl. Conference of the CLEF Association. Thessaloniki, Greece, September 22-25, 2020*, pages 255–271.

Udo Hahn, Franz Matthies, Erik Faessler, and Johannes Hellrich. 2016. UIMA-based JCORE 2.0 goes GITHUB and MAVEN CENTRAL: state-of-the-art software resource engineering and distribution of NLP pipelines. In *LREC 2016 — Proceedings of* the 10th International Conference on Language Resources and Evaluation. Portorož, Slovenia, 23-28 May 2016, pages 2502–2509.

Udo Hahn, Franz Matthies, Christina Lohr, and Markus Löffler. 2018. 3000PA : towards a national reference corpus of German clinical language. In *MIE 2018 — Proceedings of the 29th Conference on Medical Informatics in Europe. Gothenburg, Sweden, 24-26 April 2018*, pages 26–30. IOS Press.

Johannes Hellrich, Stefan Schulz, Sven Buechel, and Udo Hahn. 2015. JUFIT : a configurable rule engine for filtering and generating new multilingual UMLS terms. In *AMIA 2015 — Proceedings of the 2015 Annual Symposium of the American Medical Informatics Association. San Francisco, California, USA, Nov 14-18, 2015*, pages 604–610.

Hossein Hematialam and Wlodek W. Zadrozny. 2017. Extracting *condition-action* statements in medical guidelines. In *AMIA 2017 — Proceedings of the 2017 Annual Symposium of the American Medical Informatics Association. Washington, D.C., USA, November 4-8, 2017*.

George M. Hripcsak and Adam S. Rothschild. 2005. Agreement, the f-measure, and reliability in information retrieval. *Journal of the American Medical Informatics Association*, 12(3):296–298.

Tamseela Hussain, George Michel, and Richard N. Shiffman. 2009. The Yale Guideline Recommendation Corpus: a representative sample of the knowledge content of guidelines. *International Journal of Medical Informatics*, 78(5):354–363.

Alistair E. W. Johnson, Tom J. Pollard, Lu Shen, Liwei H. Lehman, Mengling Feng, Mohammad M. Ghassemi, Benjamin Moody, Peter Szolovits, Leo Anthony Celi, and Roger G. Mark. 2016. MIMIC-III, a freely accessible critical care database. *Scientific Data*, 3:#160035.

Katharina Kaiser, Andreas Seyfang, and Silvia Miksch. 2010. Identifying treatment activities for modelling computer-interpretable clinical practice guidelines. In *KR4HC 2010 — Selected Papers of the International Workshop on Knowledge Representation for Health Care @ ECAI 2010. Lisbon, Portugal, August 17, 2010*, pages 114–125. Springer.

Tobias Kolditz, Christina Lohr, Johannes Hellrich, Luise Modersohn, Boris Betz, Michael Kiehntopf, and Udo Hahn. 2019. Annotating German clinical documents for de-identification. In *MEDINFO 2019 — Proceedings of the 17th World Congress on Medical and Health Informatics. Lyon, France, 25-30 August 2019*, pages 203–207. IOS Press.

Maximilian König, André Sander, Ilja Demuth, Daniel Diekmann, and Elisabeth Steinhagen-Thiessen. 2019. Knowledge-based best of breed approach for automated detection of clinical events based on German free text digital hospital discharge letters. *PLoS ONE*, 14(11):#e0224916.

Jonathan Krebs, Hamo Corovic, Georg Dietrich, Maximilian Ertl, Georg Fette, Mathias Kaspar, Markus Krug, Stefan Störk, and Frank Puppe. 2017. Semi-automatic terminology generation for information extraction from German chest X-ray reports. In *GMDS 2017 — Proceedings of the 62nd Annual Meeting of the German Association of Medical Informatics, Biometry and Epidemiology. Oldenburg, Germany, 17-21 Sept. 2017*, pages 80–84. IOS Press.

Markus Kreuzthaler, Michel Oleynik, Alexander Avian, and Stefan Schulz. 2016. Unsupervised abbreviation detection in clinical narratives. In *ClinicalNLP 2016 — Proceedings of the 1st Workshop on Clinical Natural Language Processing @ COLING 2016. Osaka, Japan, December 11, 2016*, pages 91–98.

Jinhyuk Lee, Wonjin Yoon, Sungdong Kim, Donghyeon Kim, Sunkyu Kim, Chan So, and Jaewoo Kang. 2020. BIOBERT: a pre-trained biomedical language representation model for biomedical text mining. *Bioinformatics*, 36(4):1234–1240.

Tiffany I. Leung and Michel Dumontier. 2016. Overlap in drug-disease associations between clinical practice guidelines and drug structured product label indications. *Journal of Biomedical Semantics*, 7:#37.

Tiffany I. Leung, Hawre Jalal, Donna Zulman, Michel Dumontier, Douglas Owens, Mark Musen, and Mary Goldstein. 2015. Automating identification of multiple chronic conditions in clinical practice guidelines. In *Proceedings of the AMIA Joint Summits on Translational Science 2015. San Francisco, California, USA, March 23-27, 2015*, pages 456–460.

Alexander Lex, Nils Gehlenborg, Hendrik Strobelt, Romain Vuillemot, and Hanspeter Pfister. 2014. UPSET : visualization of intersecting sets. *IEEE Transactions on Visualization and Computer Graphics*, 20(12):1983–1992.

Christina Lohr, Sven Buechel, and Udo Hahn. 2018. Sharing copies of synthetic clinical corpora without physical distribution: a case study to get around IPRs and privacy constraints featuring the German JSYNCC corpus. In *LREC 2018 — Proceedings of the 11th International Conference on Language Resources and Evaluation. Miyazaki, Japan, May 7-12, 2018*, pages 1259–1266.

Christina Lohr and Robert Herms. 2016. A corpus of German clinical reports for ICD and OPS-based language modeling. In *CLAW 2016 — Proceedings of the 6th Workshop on Controlled Language Applications @ LREC 2016. Portorož, Slovenia, 28 May 2016*, pages 20–23.

Christina Lohr, Luise Modersohn, Johannes Hellrich, Tobias Kolditz, and Udo Hahn. 2020. An evolutionary approach to the annotation of discharge summaries. In *MIE 2020 — Proceedings of the 30th Conference on Medical Informatics Europe. Geneva, Switzerland, April 28 - May 1, 2020*, pages 28–32.

Mor Peleg. 2013. Computer-interpretable clinical guidelines: a methodological review. *Journal of Biomedical Informatics*, 46(4):744–763.

Jonathon L. Read, Erik Velldal, Marc Cavazza, and Gersende Georg. 2016. A corpus of clinical practice guidelines annotated with the importance of recommendations. In *LREC 2016 — Proceedings of the 10th International Conference on Language Resources and Evaluation. Portorož, Slovenia, 23-28 May 2016*, pages 1724–1731.

Roland Roller, Hans Uszkoreit, Feiyu Xu, Laura Seiffe, Michael Mikhailov, Oliver Staeck, Klemens Budde, Fabian Halleck, and Danilo Schmidt. 2016. A fine-grained corpus annotation schema of German nephrology records. In *ClinicalNLP 2016 — Proceedings of the 1st Workshop on Clinical Natural Language Processing @ COLING 2016. Osaka, Japan, December 11, 2016*, pages 69–77.

Matthieu-P. Schapranow, Milena Kraus, Cindy Perscheid, Cornelius Bock, Franz Liedke, and Hasso Plattner. 2015. The Medical Knowledge Cockpit: real-time analysis of big medical data enabling precision medicine. In *BIBM 2015 — Proceedings of the 2015 IEEE International Conference on Bioinformatics and Biomedicine. Washington, DC, USA, 9-12 November 2015*, pages 770–775.

Laura Seiffe, Oliver Marten, Michael Mikhailov, Sven Schmeier, Sebastian Möller, and Roland Roller. 2020. From witch's shot to music making bones: resources for medical laymen to technical language and vice versa. In *LREC 2020 — Proceedings of the 12th International Conference on Language Resources and Evaluation. Marseille, France, May 11-16, 2020*, pages 6185–6192.

Radu Serban, Annette ten Teije, Frank van Harmelen, Mar Marcos, and Cristina Polo-Conde. 2007. Extraction and use of linguistic patterns for modelling medical guidelines. *Artificial Intelligence in Medicine*, 39(2):137–149.

Thomas Seufferlein, Ina Kopp, Stefan Post, Walter Jonat, Rolf Kreienberg, Monika Nothacker, Annika Marks, Gerd Nettekoven, Thomas Langer, Markus Follmann, and Michael Bamberg. 2019. Onkologische Leitlinien: Herausforderungen und zukünftige Entwicklungen. *Forum*, 34:277–283.

Pontus Stenetorp, Sampo Pyysalo, Goran Topić, Tomoko Ohta, Sophia Ananiadou, and Jun'ichi Tsujii. 2012. BRAT: a Web-based tool for NLP-assisted text annotation. In *EACL 2012 — Proceedings of the 13th Conf. of the European Chapter of the Association for Computational Linguistics: Demonstrations. Avignon, France, April 25-26, 2012*, pages 102–107.

Jannik Strötgen and Michael Gertz. 2011. WIKIWARS-DE : a German corpus of narratives annotated with temporal expressions. In *GSCL 2011—Proceedings of the Conference of the German Society for Computational Linguistics and Language Technology. Hamburg, Germany, 28-30 Sept. 2011*, pages 129–134.

Jannik Strötgen, Anne-Lyse Minard, Lukas Lange, Manuela Speranza, and Bernardo Magnini. 2018. KRAUTS: a German temporally annotated news corpus. In *LREC 2018 — Proceedings of the 11th Intl. Conference on Language Resources and Evaluation. Miyazaki, Japan, May 7-12, 2018*, pages 536–540.

Maria Taboada, Maria Meizoso, Diego Martínez, David Riaño, and Albert Alonso. 2013. Combining open-source natural language processing tools to parse clinical practice guidelines. *Expert Systems*, 30(1):3–11.

Martin Toepfer, Hamo Corovic, Georg Fette, Peter Klügl, Stefan Störk, and Frank Puppe. 2015. Fine-grained information extraction from German transthoracic echocardiography reports. *BMC Medical Informatics and Decision Making*, 15:#91.

Özlem Uzuner, Brett R. South, Shuying Shen, and Scott L. DuVall. 2011. 2010 I2B2/VA Challenge on concepts, assertions, and relations in clinical text. *Journal of the American Medical Informatics Association*, 18(5):552–556.

Davy Weissenbacher, Abeed Sarker, Arjun Magge, Ashlynn R. Daughton, Karen O'Connor, Michael J. Paul, and G. Gonzalez-Hernandez. 2019. Overview of the 4th Social Media Mining for Health (#SMM-4H) Shared Task at ACL 2019. In *#SMM4H 2019 — Proceedings of the 4th Workshop on Social Media Mining for Health Applications Shared Task @ ACL 2019. Florence, Italy, August 2, 2019*, pages 21–30.

Reinhardt Wenzina and Katharina Kaiser. 2013. Identifying *condition-action* sentences using a heuristic-based information extraction method. In *KR4HC-ProHealth 2013 — Selected Papers of the Joint International Workshop on Knowledge Representation for Health Care & Process-oriented Information Systems in Healthcare @ AIME 2013. Murcia, Spain, June 1, 2013*, pages 26–38. Springer.

Joachim Wermter and Udo Hahn. 2004. Really, is medical sublanguage that different? Experimental counter-evidence from tagging medical and newspaper corpora. In *MEDINFO 2004 — Proceedings of the 11th World Congress on Medical Informatics. San Francisco, California, USA, September 7-11, 2004*, volume 1, pages 560–564. IOS Press.

Wlodek W. Zadrozny, Hossein Hematialam, and Luciana Garbayo. 2017. Towards semantic modeling of contradictions and disagreements: a case study of medical guidelines. In *IWCS 2017 — Proceedings of the 12th International Conference on Computational Semantics. Montpellier, France 19-22 September 2017*, volume 2: Short Papers, page #43.

Huijia Zhu, Yuan Ni, Peng Cai, and Feng Cao. 2013. Automatic information extraction for computerized clinical guideline. In *MEDINFO 2013 — Proceedings of the 14th World Congress on Medical and Health Informatics. Copenhagen, Denmark, 20-23 August 2013*, page 1023. IOS Press.

Normalization of Long-tail Adverse Drug Reactions in Social Media

Emmanouil Manousogiannis[2], Sepideh Mesbah[1], Alessandro Bozzon[1],
Robert-Jan Sips[2], Zoltán Szlávik[2], and Selene Báez Santamaría[2]

[1]Delft University of Technology, Landbergstraat 15, 2628 CE Delft , the Netherlands
[2]MyTomorrows, Anthony Fokkerweg 61, 1059 CP Amsterdam, the Netherlands
`manousogm@gmail.com`, {`s.mesbah, a.bozzon`} `@tudelft.nl`
`s.baezsantamaria@vu.nl`,{`r.sips, zoltan.szlavik`} `@mytomorrows.com`

Abstract

The automatic mapping of Adverse Drug Reaction (ADR) reports from user-generated content to concepts in a controlled medical vocabulary provides valuable insights for monitoring public health. While state-of-the-art deep learning-based sequence classification techniques achieve impressive performance for medical concepts with large amounts of training data, they show their limit with long-tail concepts that have a low number of training samples. The above hinders their adaptability to the changes of layman's terminology and the constant emergence of new informal medical terms. Our objective in this paper is to tackle the problem of normalizing long-tail ADR mentions in user-generated content. In this paper, we exploit the implicit semantics of rare ADRs for which we have few training samples, in order to detect the most similar class for the given ADR. The evaluation results demonstrate that our proposed approach addresses the limitations of the existing techniques when the amount of training data is limited.

1 Introduction

Discovering adverse drug reactions (ADRs) is a critical component of drug safety. In addition to controlled clinical trials, continuous monitoring of adverse effects after market introduction provides valuable insights into ADRs. Studies have shown that traditional techniques (i.e., voluntary and mandatory reporting of ADRs by patients) of post-market ADRs are not able to fully characterize drugs' adverse effects (Harpaz et al., 2012; Chee et al., 2011; Ahmad, 2003; Sarker et al., 2015).

Social media could help to obtain more information on the occurrence of adverse effects in the real world, by monitoring the information discussed and shared by users for their personal experiences with pharmaceutical drugs. Such monitoring can aid in the monitoring of public health (Aramaki et al., 2011; Paul and Dredze, 2011), and provide new opportunities for the identification of adverse drug reactions (Lee et al., 2017b; Sarker and Gonzalez, 2015; Mesbah et al., 2019).

However, web users report ADRs using a different language style and terminology that depends on the user's medical proficiency, but also on the type of online medium (e.g. health forums vs micro-post social networks). Therefore, ADR reports from user generated content typically differ significantly from ADR statements in professional medical text. As exemplified in Table 1, lay people often use diverse dialects (Karisani and Agichtein, 2018) when describing medical concepts, and make abundant use of figures of speech (e.g. metaphors) and informal terminology. Additionally, social media text is usually informal and succinct, often due to limitations imposed by the communication platform, limiting thus the extent and semantic richness of the report (Baron, 2010).

A critical step in the practical use of user-generated text for ADR surveillance starts with the normalization of reported adverse events, meaning linking the user-reported event to a formal Knowledge Base such as the UMLS (Unified Medical Language System)[1] or its subsets like MedDRA (Medical Dictionary for Regulatory Activities)[2] .

There is a large body of work on ADR normalization in social media such as *Twitter* (Chowdhury et al., 2018; Nikfarjam et al., 2015; Manousogiannis et al., 2019), or forums like *Dailystrength* (Leaman et al., 2010; Nikfarjam and Gonzalez, 2011). Existing research on ADR normalization rely on different techniques such as rule-based,

---

[1]UMLS contains structured information about a large number of different medical entities like Diseases, Symptoms, Drugs etc, as well as the relations that those different entities have with each other (for instance a certain Disease is associated with certain Symptoms)

[2]https://www.meddra.org/

49

Proceedings of the 11th International Workshop on Health Text Mining and Information Analysis, pages 49–58
November 20, 2020. ©2020 Association for Computational Linguistics
https://doi.org/10.18653/v1/P17

Layman's terminology	Medical concept in UMLS
'head spinning a little'	Dizziness
'lose 10 lbs'	Body Weight Decreased
'appetite on 10'	Increased Appetite
'terrible headache!!!!'	Headache

Table 1: Examples of layman's text describing ADRs and their related medical concept in UMLS (Unified Medical Language System)

machine translation, supervised-learning (Aronson and Lang, 2010; Stewart et al., 2012; Combi et al., 2018; Soldaini, 2016; Leaman et al., 2013; Leaman and Lu, 2014) which all showed to have limited performance compared to deep learning- based techniques (Lee et al., 2017a; Limsopatham and Collier, 2016; Tutubalina et al., 2018; Han et al., 2017; Niu et al., 2018).

Original contribution. Driven by previous literature, we note that deep neural networks demonstrate a remarkable performance in many publicly available datasets with user-generated text. However, we show that this set-up does not fully correspond to a real-world setting, where: 1) The full collection of medical concepts in the controlled vocabulary is large and continuously increasing. 2) Many concepts have limited and insufficient training examples available. This is clearly shown in Figure 1, where we can see that more than 41% of the medical concepts (classes) present in SMM4H 2017 Twitter and CADEC dataset have just one training sample. As reported in SIDER (Side Effect Resource)[3], there are more than 5000 MedDRA codes mentioned as ADRs in medical documents, but around only 10% of them (e.g., 300-500) are contained in the publicly available training data for ADR normalization. This is unrealistic in comparison to mining for ADRs 'in the wild', as the language is evolving in online and offline communication (Kershaw et al., 2016), and there is a constant emergence of new informal medical terms. This means ADR surveillance in the real world requires robustness for rare ADRs rather than performing well only in common classes. In this paper, we introduce a simple but competitive technique to adapt to the changes and emergence of informal medical layman's terms with no training costs. Our intuition is that for rare ADRs for which we have few training samples, we can exploit its implicit semantics to detect the most similar class for the given ADR. In this case, we reduce the risk of overfitting. Based on this intuition, we present a technique that leverages pre-trained language representation models and revisits a simple Nearest-Neighbor (1-NN) approach to solve the real-world problem of normalizing rare ADR concepts.

We perform an extensive experimental evaluation of our presented approach and compare its effectiveness to the current state of the art on real-world data from social media. Our evaluation aims to provide the necessary insights into the strengths and weaknesses of the proposed approach from both quantitative and qualitative perspectives. Results show that our approach achieves superior performance to state-of-the-art deep learning methods for rare ADR concepts.

2   Related Work

Existing methods for normalizing ADRs in clinical text content fall into five categories: 1) *rule-based approaches* (Aronson and Lang, 2010; Stewart et al., 2012; Combi et al., 2018; Soldaini, 2016), ; 2) *deep learning approaches* (Lee et al., 2017a; Limsopatham and Collier, 2016; Tutubalina et al., 2018; Han et al., 2017; Niu et al., 2018) 3) *machine translation techniques* (Ghiasvand and Kate, 2014; Cossin et al., 2018; Lu et al., 2017); 4) other *supervised approaches* such as DNorm (Leaman et al., 2013; Leaman and Lu, 2014) and 5) *unsupervised techniques* (Tahmasebi et al., 2018).

Rule-based approaches mostly rely on string matching techniques, ignoring the semantics of each text mention, which results in a relatively poor performance (Limsopatham and Collier, 2016). The *unsupervised* techniques, despite the fact that they capture semantic similarity between text with minimal string similarity, barely outperform the rule-based techniques. The use of language in domains like social media or online forums is totally different than the official medical terminology used in Knowledge Bases and hence it is hard to find

---

[3] http://sideeffects.embl.de/

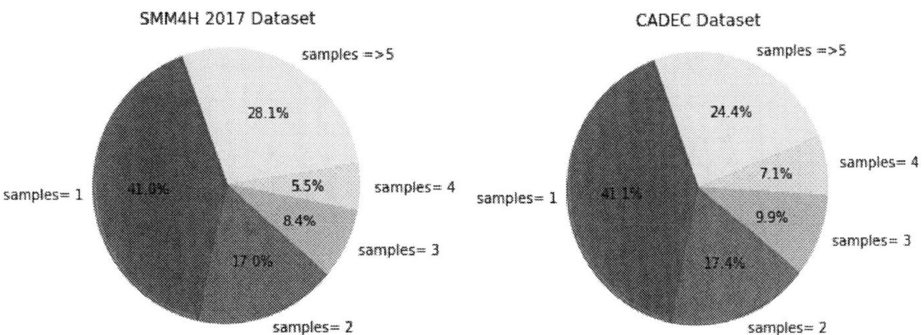

Figure 1: Available training samples per concept in
SMM4H 2017 Twitter and CADEC Dataset

a common embedding space that is able to match semantically similar entities. The results published in (Limsopatham and Collier, 2016; Lee et al., 2017a; Tutubalina et al., 2018; Niu et al., 2018) show that supervised machine learning techniques such as *deep learning* and *machine translation* clearly outperform rule-based and unsupervised techniques (i.e., deep learning achieving the highest accuracy). In addition to the task specific deep learning techniques, BERT (Devlin et al., 2018) a bi-directional language representation model was recently fine-tuned in order to tackle the ADR normalization problem as a sequence classification task . This attempt demonstrated remarkable performance across different user-generated text datasets (Miftakhutdinov and Tutubalina, 2019). However, to perform properly, supervised techniques require large collection of labeled training data for each concept and are not suitable for normalizing concepts with none or few training samples. This is clearly shown in Figure 2, were we reproduced the state of the art RNN as presented in (Limsopatham and Collier, 2016) and also fine-tuned BERT, to visualize the performance of those models as a function of the available training samples that each concept (class) has on the SMM4H 2017 Twitter Dataset.

In social media posts some medical concepts like ADRs or symptoms, are way more common than others. For instance, it is common to find many different expressions referring to 'Headache' or 'Stomach Pain' caused by a drug use, rather than 'sleepwalking'. As a result, there is a high class imbalance between common and rare medical concepts that usually show up in social media. In Figure 1, we can clearly see that more than 40 % of the medical concepts (classes) present in this dataset (507) have just one training sample. In this context, it becomes obvious that there is a need for an alternative way of predicting concepts when we have insufficient training data.

Addressing the problem of scarcity in training data is a well-studied research topic. Few-shot learning (Wang and Yao, 2019) is a family of machine learning algorithms that are able to perform classification with only a few 'shots' (samples) from each class. To achieve that, most of those techniques leverage less complex models to avoid overfitting on the limited amount of training data. For instance, embedding-based few-shot learning models (Wang and Yao, 2019), try to classify unknown samples by creating a meaningful embedding (feature representation) of the labeled and unlabeled data and then classifying the unlabeled samples based on their most similar embedding from the training data. Recent studies (Wang et al., 2019) show that this family of Nearest Neighbor based approaches can achieve surprisingly promising results in a variety of tasks.

In Natural Language Processing, this family of techniques can profit from various word embedding models (Pennington et al., 2014; Mikolov et al., 2013; Bojanowski et al., 2017) to create a semantically meaningful representation of text. Those embedding models are proven to create vector representation of words that can capture their semantics, in such a way that semantically similar words will have similar vectors, in terms of cosine similarity. On the sentence and phrase level, several techniques can be used to derive a fixed-size vector representation for the whole phrase, regardless of the number of tokens the phrase includes. These techniques can vary from simple approaches like calculating the average of the individual token em-

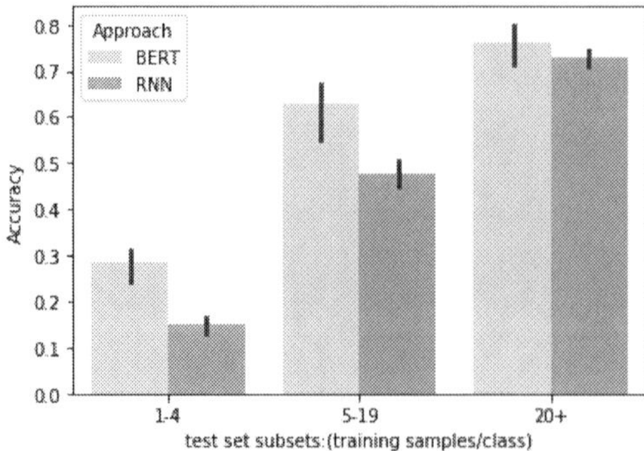

Figure 2: Accuracy of RNN and BERT as a function of the available training samples on SMM4H 2017 dataset.

beddings of a phrase, to more sophisticated techniques that use encoder-based models to derive a meaningful fixed-size vector for each multi-token phrase they receive as input. Recently, researchers in SBERT (Reimers and Gurevych, 2019) modified the BERT language representation model, leveraging Siamese and Triplet network structures, to derive semantically meaningful sentence embeddings that can be compared using cosine-similarity. The above was demonstrated as a primary drawback of the original BERT version. The network was fine-tuned with pairs of sentences from Natural Language Inference (NLI) and Semantic Textual Similarity (STS) datasets achieving state of the art performance in minimizing the embedding distance between semantically similar sentences compared to previous sentence embedding techniques.

Despite the aforementioned research in this domain, to our knowledge, none of the previous research in medical concept normalization in user-generated text proposed a solution that would address the problem of class imbalance and scarcity in the training data.

3   ADR Normalization

In the medical domain, getting more annotated data for rare concepts would be extremely expensive and time consuming. Therefore we need to focus our effort on predicting a class (medical concept) for which we have none or maybe few representatives. On a theoretical level, our basic hypothesis is to test whether creating fixed-size vector representations of ADRs from the training and test data is a valuable feature in order to normalize the unlabeled ADR mentions, based only on their vector similarities. In that case, we are expecting that a

Nearest Neighbor (1-NN) approach can take advantage of its simplicity and demonstrate a better performance in medical concepts where training data is limited. As we mentioned before, this is a significant percentage of the data in a real world scenario.

Our method is composed of the following stages: 1) we use sentence embedding techniques to encode all ADR phrases in the training and test data into a fixed-size vector so that similar ADRs appear in close proximity; 2) we use cosine distance to measure the semantic similarity between the vector representation of an unlabeled ADR and each representation of the training samples; 3) we classify the unlabeled ADR with the label of its Nearest Neighbor from the training data. There are different techniques to encode words or phrases into a fixed-size vector representation. In this paper, we use two techniques, a simple averaging approach and the state-of-the-art sentence representation model (Reimers and Gurevych, 2019) described below:

- Simple average of word embeddings (avg-wordpiece): We create a fixed-size vector representation of an ADR phrase by using a pre-trained word embedding model, encoding each token in an ADR phrase and then averaging the individual token embeddings. As a pre-trained word embedding model, we use WordPiece embeddings used by BERT (Devlin et al., 2018) in order to establish a fair comparison with the BERT fine-tuned model.

- S-BERT encoder (SBERT): We also used Sentence BERT (Reimers and Gurevych, 2019), which is the state-of-the-art sentence representation model for semantic textual similarity tasks. Its Siamese network architecture is composed of two identical pre-trained BERT models. This model is then trained on pairs of similar and dissimilar sentences with the objective to reduce the distance between the semantically similar pairs and increase the distance between the dissimilar pairs. Further details can be found in the original implementation of this model.

SBERT model training details: To train the model for our task, we needed pairs of similar and dissimilar ADR phrases, which is a time-consuming and expensive process to obtain. For this reason, we used the two pre-trained models provided by the authors. The first model is trained on a combination of the SNLI (Bowman et al., 2015) and MNLI (Williams et al., 2017) datasets, consisting of approximately 1 million sentence pairs labeled as a contradiction, entailment, or neutral. The second one is trained on the aforementioned NLI data, and then further tuned on the STSb dataset (Tian et al., 2017), which consists of 8,628 sentence pairs from image captions and news. Those pairs are labeled with a number between 0 and 5, indicating the semantic relatedness of each sentence pair. Besides, we also experimented with medNLI (Romanov and Shivade, 2018) and BIOSSES STS (Soğancıoğlu et al., 2017) datasets, which are manually annotated sentence pairs on the clinical domain. Despite the different combinations of data tried, the best performing version was the provided pre-trained model on SNLI, MNLI, and STSb data, which will be further used in our experiments.

After encoding the phrases from the train and test data in the semantic vector space, the normalization of the test samples is done based on their nearest neighbor from the training data. As a similarity measure, we use cosine similarity, which is the most common similarity measure between word vector representations (Lofi, 2015). We finally assign the concept label of the Nearest Neighbor (i.e., the sample with the highest similarity measure) from the training data to the unlabeled ADR in the test set.

4 Experimental Settings

As an evaluation metric we use accuracy, which denotes the percentage of the correctly normalized ADR samples in the test data. The focus of our evaluation is on the variation of performance using the complete set of training data as well as the different fractions of available training data per class. Those subsets are selected based on how many training samples the true label of a test set sample has in the training data. In this way we can demonstrate the effectiveness of our technique in dealing with rare ADRs. Furthermore, we include the results of qualitative analysis, which will help us identify in more detail the strengths and the limitations of the proposed approach

4.1 Dataset Description

We evaluated our approach on SMM4H 2017 (Sarker et al., 2018) (i.e., the largest available Twitter dataset to our knowledge) and the Cadec dataset (Karimi et al., 2015). **Twitter SMM4H 2017** was published as part of the Social Media Mining for

Dataset	#ADR mentions	#MEDDRA Codes
SMM4H 2017	3629	507
CADEC	3092	659

Table 2: Summary of Datasets used in the experimental procedure

Health workshop in 2017. Unfortunately, the annotation data was only released for the training set and the development set. The annotations for the test set were not released in public. For this reason, we use the annotated part of the data for our experiments. The development set and train set were originally concatenated and then split into **5 equal folds**. However, out of the approximately 9500 mentions only **3629 ADR mentions** were unique. Those mentions were mapped to **507 medical concepts** from the MEDDRA Knowledge-Base. As there was a very high percentage of overlap between the training and test folds, we decided to remove all duplicates in order to avoid over optimistic estimations of our performance. After the duplicate removal, from the validation (development) and test set folds, the final test sets consisted of approximately 400 ADR mentions for testing and 3200 mentions for training in each one of the 5 different folds.

Apart from SMM4H 2017 dataset, we leveraged **CADEC** dataset. This data does not only consist of concepts representing Adverse Drug Reactions. It also includes other medical concepts like drugs, diseases and symptoms. For this reason, we filtered all the annotated samples and excluded all data samples that represented different medical concepts, based on their assigned MedDRA code. The resulting dataset we used for evaluation consisted of **3092 samples** mapped to **659 distinct MedDRA codes**. Statistics on data used for training and testing are shown in Table 2.

4.2 Comparison Methods

In order to enable a direct comparison of our 1-NN approach, we reproduced two state of the art neural networks, **BERT** (Miftakhutdinov and Tutubalina, 2019) and the second is a **Recurrent Neural Network (RNN)** (Limsopatham and Collier, 2016) with a single GRU layer. In (Miftakhutdinov and Tutubalina, 2019), the authors fine-tuned BERT for ADR sequence classification, which demonstrated a remarkable performance and outperformed all task specific neural network architectures. In our experiments we are using the BERT-base version

Approach	SMM4H 2017	CADEC
RNN	0.561	0.459
1-NN-avgwordpiece	0.587	0.484
BERT	**0.667**	**0.571**
1-NN-SBERT	0.637	0.531

Table 3: 5 fold cross validation accuracy on SMM4H 2017 and CADEC datasets. Legend: 1-NN-avgwordpiece – using Nearest Neighbor (NN) with the avgwordpiece representation model; 1-NN-SBERT – using NN with SBERT representation model.

Approach	Rare classes Subset* CADEC (115 samples/fold)	Rare classes Subset* SMM4H 2017 (80 samples/fold)
BERT	0.295	0.285
1-NN-SBERT	**0.34**	**0.37**

Table 4: 5 fold cross validation accuracy of the two best performing models on a subset of the test data. The subset includes samples from classes that have less than 5 training samples in the training data.

of this model. The RNN model (Limsopatham and Collier, 2016), demonstrated superior performance compared to all other rule-based or ML based techniques in normalising medical concepts. In addition the authors made their implementation details public to the research community in order to ensure the accurate reproducibility of their approach.

5 Results And Discussion

The results of our experimental evaluation in the two aforementioned datasets are demonstrated in Table 3. Overall, BERT outperforms both the RNN and the 1-NN based models in SMM4H 2017 and CADEC datasets. However, 1-NN-SBERT (i.e., using NN with SBERT representation model) achieves a comparable performance to it, while outperforming the RNN architecture as well as the baseline 1-NN-avgwordpiece. Based on the above results, we could conclude that the BERT sequence classification technique is the big winner in the ADR normalization task, which is also in line with previous research findings (Miftakhutdinov and Tutubalina, 2019). However, as Figure 1 indicates, having insights on different subsets of the predicted medical concepts would give us a better picture of the strengths and the weaknesses of each technique.

For this reason, we divided all test sets into three different subsets, based on how many training samples the true label of test set sample has in the training data. The first test subset includes ADRs with a true label that has just 1 to 4 unique training samples in the corresponding training data. Accordingly, the second subset includes concepts with 5-19 samples in the training set and the third 20 or more of them. The bin limits were selected so that each subset has a significant number of samples.

The results of the accuracy in those subsets are shown in Figures 3 and 4. From these results, we can see that our proposed 1-NN-SBERT technique outperforms BERT and RNN in predicting classes with limited training samples (1-4). This represents the vast majority of the different medical concepts that are present in those two datasets, as seen in Figure 1. The exact accuracy metrics of the two best performing techniques in this subset are presented in Table 4. As expected, we can see that as the availability of training data increases in the other two bins, the sequence classification performance techniques are becoming more effective than the 1-NN based approach.

6 Qualitative analysis

We perform a qualitative analysis of our 1-NN-SBERT approach performance on the SMM4H 2017 dataset. The purpose of the qualitative analysis is to get insights about 'where' and 'why' this approach fails or succeeds and to highlight the underlying properties which are hard to digitize without losing any meaning. Table 5 shows examples of correctly normalized ADRs and their corresponding medical terms in the knowledge base, as well as different kinds of misclassified ADR samples. Based on our error analysis, we identified three types of errors where our approach produces the majority of the incorrectly normalized samples:

- 1-NN-SBERT fails to take into account some significant properties of textual data, like negation. The phrase 'never going to lose weight' is erroneously normalized to Weight Decrease because its closest neighbor in the vector space is 'lose so much weight'.

- There exist inconsistencies or disagreement in the manual annotation of the test data. Nonetheless, our model selected semantically similar concepts in the classification procedure. The phrase 'kills my sex drive' is normalized to the medical concept 'Loss of Li-

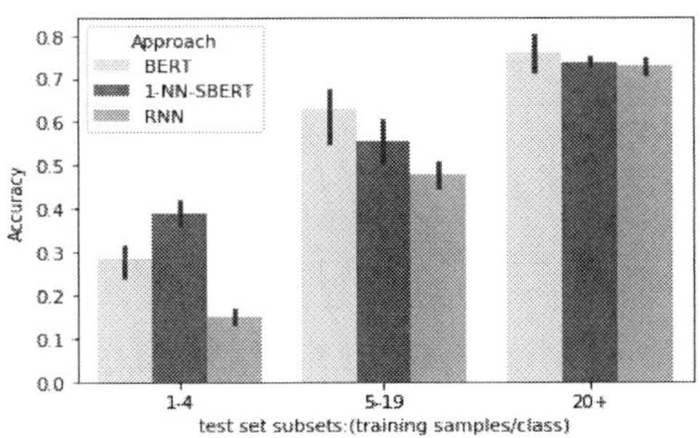

Figure 3: Accuracy of different techniques per available training samples on SMM4H 2017.

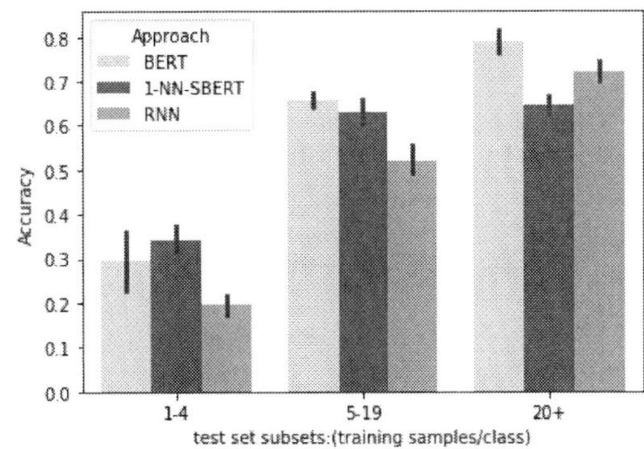

Figure 4: Accuracy of different techniques per available training samples on CADEC test set.

bido' while the ground truth is 'Libido decreased'. Despite having almost the same meaning, these medical concepts represent two different entities in MedDRA. This indicates that we could consider the medical concept normalization problem as a multi-label classification task.

• Our approach does not consider the context around an extracted ADR entity in order to achieve a more robust normalization performance. For instance, the phrase 'feel like I am having a heart attack' was once annotated as a 'Palpitations' adverse effect, while an almost identical phrase was annotated as 'Myocardial infarction'. Most likely, the annotations were based on the information provided by the rest of the text.

While the main focus of the paper was on normalizing rare concepts that have a few training samples available, we also started investigating an additional source of knowledge for the concepts for which no training data is available. For the medical concepts with no training data, we used all different synonymous terms (i.e., available in UMLS) associated with this concept as its representatives.

7 Conclusions and Future Work

In this work, we have presented a 1-NN-SBERT approach that is robust for normalizing rare classes of ADRs. Our technique cannot compete with a deep neural network in concepts where the training data is available on a larger scale. Yet, it easily outperforms deep learning techniques when dealing with rare classes of ADRs. This approach can easily scale as the number of classes increases and can

adapt to the changes and emergence of new ADRs, with no training cost. Finally, our sentence embedding model can benefit from the concept of transfer learning, as it does not require task-specific training data to generate high-quality sentence and phrase representations. As future work our model can be improved in the direction of separating the common and the rare medical concepts at test time in a more effective and efficient way. Using multiple binary classifiers, or neural networks with multiple sigmoid functions instead of the final softmax layer have been used in similar domains where multi-label classification or open-set classification (Shu et al., 2017) is considered.

References

Syed Rizwanuddin Ahmad. 2003. Adverse drug event monitoring at the Food and Drug Administration: your report can make a difference. *Journal of General Internal Medicine* 18, 1 (2003), 57–60.

Eiji Aramaki, Sachiko Maskawa, and Mizuki Morita. 2011. Twitter catches the flu: detecting influenza epidemics using Twitter. In *Proceedings of the conference on empirical methods in natural language processing*. Association for Computational Linguistics, 1568–1576.

Alan R. Aronson and François Michel Lang. 2010. An overview of MetaMap: Historical perspective and recent advances. *Journal of the American Medical Informatics Association* 17, 3 (2010), 229–236. https://doi.org/10.1136/jamia.2009.002733 arXiv:arXiv:1502.05814v1

Naomi S Baron. 2010. *Always on: Language in an online and mobile world*. Oxford University Press.

Piotr Bojanowski, Edouard Grave, Armand Joulin, and Tomas Mikolov. 2017. Enriching Word Vectors

ADR mention	Predicted Concept	Ground Truth
'Food doesn't look appetizing'	Decreased Appetite	Decreased Appetite
'Slept 10-14 hours '	Hypersomnia	Hypersomnia
'kills my sex drive'	Loss of libido	Libido decreased
'Never going to lose weight'	Weight decreased	Weight increased
'like I am having a heart attack'	Palpitations	Myocardial infarction
'fade into sleep'	Somnolence	Hypersomnia

Table 5: Examples of correctly and incorrectly normalized ADRs from our 1-NN-SBERT model.

with Subword Information. *Transactions of the Association for Computational Linguistics* 5 (2017), 135–146. https://doi.org/10.1162/tacl_a_00051

Samuel R Bowman, Gabor Angeli, Christopher Potts, and Christopher D Manning. 2015. A large annotated corpus for learning natural language inference. *arXiv preprint arXiv:1508.05326* (2015).

Brant W Chee, Richard Berlin, and Bruce Schatz. 2011. Predicting adverse drug events from personal health messages. In *AMIA Annual Symposium Proceedings*, Vol. 2011. American Medical Informatics Association, 217.

Shaika Chowdhury, Chenwei Zhang, and Philip S Yu. 2018. Multi-Task Pharmacovigilance Mining from Social Media Posts. In *Proceedings of the 27th International Conference on World Wide Web*. International World Wide Web Conferences Steering Committee, 117–126.

C Combi, M Zorzi, G Pozzani, E Arzenton, and U Moretti. 2018. Normalizing Spontaneous Reports into MedDRA: some Experiments with MagiCoder. *IEEE Journal of Biomedical and Health Informatics* (2018). https://doi.org/10.1109/JBHI.2018.2861213

Sébastien Cossin, Vianney Jouhet, Fleur Mougin, Gayo Diallo, and Frantz Thiessard. 2018. IAM at CLEF eHealth 2018: Concept annotation and coding in French death certificates. *CEUR Workshop Proceedings* 2125 (2018). arXiv:1807.03674

Jacob Devlin, Ming-Wei Chang, Kenton Lee, and Kristina Toutanova. 2018. BERT: Pre-training of Deep Bidirectional Transformers for Language Understanding. *arXiv preprint arXiv:1810.04805* (2018).

Omid Ghiasvand and Rohit J Kate. 2014. UWM : Disorder Mention Extraction from Clinical Text Using CRFs and Normalization Using Learned Edit Distance Patterns. SemEval (2014), 828–832.

Sifei Han, Tung Tran, Anthony Rios, and Ramakanth Kavuluru. 2017. Team UKNLP: Detecting ADRs, Classifying Medication Intake Messages, and Normalizing ADR Mentions on Twitter. *Proceedings of the 2nd Workshop on Social Media Mining for*

Health Research and Applications (2017), 49–53. http://ceur-ws.org/Vol-1996/paper9.pdf

Rave Harpaz, William DuMouchel, Nigam H Shah, David Madigan, Patrick Ryan, and Carol Friedman. 2012. Novel data-mining methodologies for adverse drug event discovery and analysis. *Clinical Pharmacology & Therapeutics* 91, 6 (2012), 1010–1021.

Sarvnaz Karimi, Alejandro Metke-Jimenez, Madonna Kemp, and Chen Wang. 2015. Cadec: A corpus of adverse drug event annotations. *Journal of Biomedical Informatics* 55 (2015), 73–81.

Payam Karisani and Eugene Agichtein. 2018. Did You Really Just Have a Heart Attack?: Towards Robust Detection of Personal Health Mentions in Social Media. In *Proceedings of the 2018 World Wide Web Conference on World Wide Web*. International World Wide Web Conferences Steering Committee, 137–146.

Daniel Kershaw, Matthew Rowe, and Patrick Stacey. 2016. Towards modelling language innovation acceptance in online social networks. In *Proceedings of the Ninth ACM International Conference on Web Search and Data Mining*. ACM, 553–562.

Robert Leaman, Rezarta Islamaj Doğan, and Zhiyong Lu. 2013. DNorm: Disease name normalization with pairwise learning to rank. *Bioinformatics* 29, 22 (2013), 2909–2917. https://doi.org/10.1093/bioinformatics/btt474

Robert Leaman and Zhiyong Lu. 2014. Disease Named Entity Recognition and Normalization with DNorm. In *Proceedings of the 5th ACM Conference on Bioinformatics, Computational Biology, and Health Informatics (BCB '14)*. ACM, New York, NY, USA, 587. https://doi.org/10.1145/2649387.2660780

Robert Leaman, Laura Wojtulewicz, Ryan Sullivan, Annie Skariah, Jian Yang, and Graciela Gonzalez. 2010. Towards internet-age pharmacovigilance: extracting adverse drug reactions from user posts to health-related social networks. In *Proceedings of the 2010 workshop on biomedical natural language processing*. Association for Computational Linguistics, 117–125.

Kathy Lee, Sadid A. Hasan, Oladimeji Farri, Alok Choudhary, and Ankit Agrawal. 2017a. Medical

Concept Normalization for Online User-Generated Texts. *Proceedings - 2017 IEEE International Conference on Healthcare Informatics, ICHI 2017* (aug 2017), 462–469. https://doi.org/10.1109/ICHI.2017.59

Kathy Lee, Ashequl Qadir, Sadid A Hasan, Vivek Datla, Aaditya Prakash, Joey Liu, and Oladimeji Farri. 2017b. Adverse drug event detection in tweets with semi-supervised convolutional neural networks. In *Proceedings of the 26th International Conference on World Wide Web*. International World Wide Web Conferences Steering Committee, 705–714.

N Limsopatham and N Collier. 2016. Normalising medical concepts in social media texts by learning semantic representation. In *54th Annual Meeting of the Association for Computational Linguistics, ACL 2016 - Long Papers*, Vol. 2. 1014–1023. https://www.scopus.com/inward/record.uri?eid=2-s2.0-85011838424{&}partnerID=40{&}md5=accab69c7d3d0bb0bf95a93ee0415605

Christoph Lofi. 2015. Measuring semantic similarity and relatedness with distributional and knowledge-based approaches. *Information and Media Technologies* 10, 3 (2015), 493–501.

C.J. Lu, D. Tormey, L. McCreedy, and A.C. Browne. 2017. *Enhanced lexsynonym acquisition for effective UMLS concept mapping*. Vol. 245. 501–505 pages. https://doi.org/10.3233/978-1-61499-830-3-501

Emmanouil Manousogiannis, Sepideh Mesbah, Alessandro Bozzon, Selene Baez, and Robert Jan Sips. 2019. Give it a shot: Few-shot learning to normalize ADR mentions in Social Media posts. In *Proceedings of the Fourth Social Media Mining for Health Applications (# SMM4H) Workshop & Shared Task*. 114–116.

Sepideh Mesbah, Jie Yang, Robert-Jan Sips, Manuel Valle Torre, Christoph Lofi, Alessandro Bozzon, and Geert-Jan Houben. 2019. Training Data Augmentation for Detecting Adverse Drug Reactions in User-Generated Content. In *Proceedings of the 2019 Conference on Empirical Methods in Natural Language Processing and the 9th International Joint Conference on Natural Language Processing (EMNLP-IJCNLP)*. Association for Computational Linguistics, Hong Kong, China, 2349–2359. https://doi.org/10.18653/v1/D19-1239

Zulfat Miftakhutdinov and Elena Tutubalina. 2019. Deep Neural Models for Medical Concept Normalization in User-Generated Texts. https://doi.org/10.18653/v1/P19-2055

Tomas Mikolov, Kai Chen, G.s Corrado, and Jeffrey Dean. 2013. Efficient Estimation of Word Representations in Vector Space. *Proceedings of Workshop at ICLR* 2013 (01 2013).

Azadeh Nikfarjam and Graciela H Gonzalez. 2011. Pattern mining for extraction of mentions of adverse drug reactions from user comments. In *AMIA Annual Symposium Proceedings*, Vol. 2011. American Medical Informatics Association, 1019.

Azadeh Nikfarjam, Abeed Sarker, Karen O'connor, Rachel Ginn, and Graciela Gonzalez. 2015. Pharmacovigilance from social media: mining adverse drug reaction mentions using sequence labeling with word embedding cluster features. *Journal of the American Medical Informatics Association* 22, 3 (2015), 671–681.

Jinghao Niu, Yehui Yang, Siheng Zhang, Zhengya Sun, and Wensheng Zhang. 2018. Multi-task Character-Level Attentional Networks for Medical Concept Normalization. *Neural Processing Letters* (2018), 1–18. https://doi.org/10.1007/s11063-018-9873-x arXiv:1005.4198

Michael J Paul and Mark Dredze. 2011. You are what you Tweet: Analyzing Twitter for public health. *Icwsm* 20 (2011), 265–272.

Jeffrey Pennington, Richard Socher, and Christoper Manning. 2014. Glove: Global Vectors for Word Representation. *EMNLP* 14, 1532–1543. https://doi.org/10.3115/v1/D14-1162

Nils Reimers and Iryna Gurevych. 2019. Sentence-bert: Sentence embeddings using siamese bert-networks. *arXiv preprint arXiv:1908.10084* (2019).

Alexey Romanov and Chaitanya Shivade. 2018. Lessons from natural language inference in the clinical domain. *arXiv preprint arXiv:1808.06752* (2018).

Abeed Sarker, Maksim Belousov, Jasper Friedrichs, Kai Hakala, Svetlana Kiritchenko, Farrokh Mehryary, Sifei Han, Tung Tran, Anthony Rios, Ramakanth Kavuluru, et al. 2018. Data and systems for medication-related text classification and concept normalization from Twitter: insights from the Social Media Mining for Health (SMM4H)-2017 shared task. *Journal of the American Medical Informatics Association* 25, 10 (2018), 1274–1283.

Abeed Sarker, Rachel Ginn, Azadeh Nikfarjam, Karen O'Connor, Karen Smith, Swetha Jayaraman, Tejaswi Upadhaya, and Graciela Gonzalez. 2015. Utilizing social media data for pharmacovigilance: a review. *Journal of Biomedical Informatics* 54 (2015), 202–212.

Abeed Sarker and Graciela Gonzalez. 2015. Portable automatic text classification for adverse drug reaction detection via multi-corpus training. *Journal of Biomedical Informatics* 53 (2015), 196–207.

Lei Shu, Hu Xu, and Bing Liu. 2017. DOC: Deep Open Classification of Text Documents. In *Proceedings of the 2017 Conference on Empirical Methods in Natural Language Processing*. Association for Computational Linguistics, Copenhagen, Denmark,

2911–2916. `https://doi.org/10.18653/v1/D17-1314`

Gizem Soğancıoğlu, Hakime Öztürk, and Arzucan Özgür. 2017. BIOSSES: a semantic sentence similarity estimation system for the biomedical domain. *Bioinformatics* 33, 14 (2017), i49–i58.

Luca Soldaini. 2016. QuickUMLS : a fast , unsupervised approach for medical concept extraction. (2016).

S.A. Stewart, M.E. Von Maltzahn, and S.S.R. Abidi. 2012. Comparing metamap to mgrep as a tool for mapping free text to formal medical lexicons. In *CEUR Workshop Proceedings*, Vol. 895. 63–77.

Amir M Tahmasebi, Henghui Zhu, Gabriel Mankovich, Peter Prinsen, Prescott Klassen, Sam Pilato, Rob Van Ommering, Pritesh Patel, Martin L Gunn, and Paul Chang. 2018. Automatic Normalization of Anatomical Phrases in Radiology Reports Using Unsupervised Learning. (2018).

Junfeng Tian, Zhiheng Zhou, Man Lan, and Yuanbin Wu. 2017. Ecnu at semeval-2017 task 1: Leverage kernel-based traditional nlp features and neural networks to build a universal model for multilingual and cross-lingual semantic textual similarity. In *Proceedings of the 11th International Workshop on Semantic Evaluation (SemEval-2017)*. 191–197.

Elena Tutubalina, Zulfat Miftahutdinov, Sergey Nikolenko, and Valentin Malykh. 2018. Medical concept normalization in social media posts with recurrent neural networks. *Journal of Biomedical Informatics* 84, June (2018), 93–102. `https://doi.org/10.1016/j.jbi.2018.06.006`

Yan Wang, Wei-Lun Chao, Kilian Q Weinberger, and Laurens van der Maaten. 2019. SimpleShot: Revisiting Nearest-Neighbor Classification for Few-Shot Learning. *arXiv preprint arXiv:1911.04623* (2019).

Yaqing Wang and Quanming Yao. 2019. Few-shot Learning: A Survey.

Adina Williams, Nikita Nangia, and Samuel R Bowman. 2017. A broad-coverage challenge corpus for sentence understanding through inference. *arXiv preprint arXiv:1704.05426* (2017).

Evaluation of Machine Translation Methods applied to Medical Terminologies

Konstantinos Skianis
BLUAI
Athens, Greece
skianis.konstantinos@gmail.com

Yann Briand, Florent Desgrippes
Agence du Numérique en Santé
Paris, France
yann.briand@sante.gouv.fr
florent.desgrippes@gmail.com

Abstract

Medical terminologies resources and standards play vital roles in clinical data exchanges, enabling significantly the services' interoperability within healthcare national information networks. Health and medical science are constantly evolving causing requirements to advance the terminologies editions. In this paper, we present our evaluation work of the latest machine translation techniques addressing medical terminologies. Experiments have been conducted leveraging selected statistical and neural machine translation methods. The devised procedure is tested on a validated sample of ICD-11 and ICF terminologies from English to French with promising results.

1 Introduction

Medical terminologies are of essential importance for health institutions to store, organize and exchange all medical-related data generated in labs, hospitals and other healthcare entities. They are arranged systematically in dictionaries and lexicons, that follow specific structures and coding rules. In order to facilitate hierarchies and connections, the terms are represented by ontologies, enabling us to keep additional information (e.g. a family of diseases).

WHO International Classification of Diseases (ICD)[1] terminology is a diagnostic classification standard for epidemiology, clinical and research purposes. It is the most used medical dictionary across national health organizations worldwide. WHO is responsible to maintain the ICD editions for the English language. ICD-11 is the latest edition, adopted on May 25th, 2019. As the initial medical lexicons which contain these ontologies are created in English, there is an evident need for translation in other languages. This translation

process can be expensive both in terms of time and resources, while the vocabulary and number of medical terms can reach high numbers and require health professional efforts for evaluation.

This work constitutes a generic, language-independent and open methodology for medical terminology translation. To illustrate our approach, which is based on automated machine translation methods, we will attempt to develop a first baseline translation from English to French for the ICD-11 classification. We also test on the International Classification of Functioning, Disability and Health (ICF) terminology[2].

First, we are going to investigate existing machine translation research studies concerning medical terms and documents, with a comparison of the relative methods. Next, we present our proposed methodology. Afterwards, we show our experiments and results. Last, we conclude with recommendations for future work.

2 Related Work

Translating medical terminologies has been a well-studied topic, with many approaches coming from machine translation. Traditional machine translation models first incorporated statistical models, whose parameters are set through the analysis of bilingual text corpora.

Statistical machine translation (SMT) Eck et al. (2004) investigated the usefulness of a large medical database (the Unified Medical Language System) for the translation of dialogues between doctors and patients using a statistical machine translation system. They showed that the extraction of a large dictionary and the usage of semantic type information to generalize the training data significantly improves the translation performance.

---

[1] https://icd.who.int/en

[2] http://bioportal.lirmm.fr/ontologies/ICF

59

Proceedings of the 11th International Workshop on Health Text Mining and Information Analysis, pages 59–69
November 20, 2020. ©2020 Association for Computational Linguistics
https://doi.org/10.18653/v1/P17

	Resources	Type	Method	Languages
Nyström et al. (2006)	ICD-10, ICF, MeSH	SMT	Alignment	En-Swe
Deléger et al. (2010)	MeSH, SNMI, MedDRA 17, WHO-ART	SMT	Knowledge, Corpus	En-Fr
Laroche and Langlais (2010)	Wiki	SMT	Projection-based	Fr-En
Dušek et al. (2014)	EMEA, UMLS, MAREC	SMT	Domain	Multi
Silva et al. (2015)	SNOMED CT, DBPedia	Auto	Alignment	En-Por
Wołk and Marasek (2015)	EMEA	NMT	Encoder-Decoder	Pol-En
Arcan et al. (2016)	Organic.Lingua	SMT	Domain	En-(Ge, It, Sp)
Arcan and Buitelaar (2017)	ICD, Wiki	Both	Knowledge Base	En-Ge
Renato et al. (2018)	DeCS, Dicionario Medico, Wiki	SMT	Domain	Sp-Por
Khan et al. (2018)	UFAL, PatTR	NMT	Domain	En-Fr

Table 1: Summary of recent techniques for medical terms and texts translation.

Claveau and Zweigenbaum (2005) presented a method to automatically translate a large class of terms in the biomedical domain from one language to another; it is evaluated on translations between French and English. Their technique relies on a supervised machine-learning algorithm, called OS-TIA (Oncina, 1991), that infers transducers from examples of bilingual term-pairs. Such transducers, when given a new term in English (respectively French), must propose the corresponding French (resp. English) term.

Later, Nyström et al. (2006) reports on a parallel collection of rubrics from the medical terminology systems ICD-10, ICF, MeSH, NCSP and KSH97-P and its use for semi-automatic creation of an English-Swedish dictionary of medical terminology. The methods presented are relevant for many other West European language pairs.

Deléger et al. (2009) presented a methodology aiming to ease this process by automatically acquiring new translations of medical terms based on word alignment in parallel text corpora, and test it on English and French. After collecting a parallel, English-French corpus, French translations of English terms were detected from three terminologies-MeSH, Snomed CT and the MedlinePlus Health Topics. A sample of the MeSH translations was submitted to expert review and a relatively high percentage of 61.5% were deemed desirable additions to the French MeSH. In conclusion, they successfully obtained good quality new translations, which underlines the suitability of using alignment in text corpora to help translating terminologies. Their method may be applied to different European languages and provides a methodological framework that may be used with different processing tools.

Neural machine translation (NMT) In recent years, NMT has emerged as the state-of-the-art approach. NMT uses a large artificial neural network which takes as an input a source sentence $(x_1, \ldots, x_m)$ and generates its translation $(y_1, \ldots, y_n)$, where x and y are source and target words respectively. Till recently, the dominant approach to NMT encodes the input sequence and subsequently generates a variable length translated sequence using recurrent neural networks (RNN) (Bahdanau et al., 2014; Sutskever et al., 2014). NMT differs entirely from phrase-based statistical approaches that use separately engineered subcomponents (Wołk and Marasek, 2015).

Domain adaptation In machine translation, domain adaptation can be applied when a large amount of out-of-domain data co-exists with a small amount of in-domain data.

Arcan and Buitelaar (2017) presented a performance comparison between SMT and NMT methods on translating highly domain-specific expressions, i.e. terminologies, documented in the ICD ontology from the medical domain. They showed that domain adaptation with only terminological expressions significantly improves the translation quality, which is specifically evident if an existing generic neural network is retrained with a limited vocabulary of the targeted domain. Last, they observed the benefit of subword models over word-based NMT models for terminology translation.

All previous work focus on training with specific terminologies. Although these methods are widely used, their vocabulary may be limited. Moreover, their size is not sufficient for training NMT methods, resulting in low translation performance.

To address these problems, Khan et al. (2018) trained NMT systems by applying transfer learning. Transfer learning falls under the umbrella of domain adaptation. In transfer learning the knowledge learned from a pre-trained existing model is

Terminology	Size	avg_len(en)	Incl	avg_len(en)
ICD-10	32474	5.49	7655	3.78
CHU Rouen HeTOP	202402	3.63	3892	3.69
ORDO	50425	6.2	3716	5.56
ACAD	47603	2.45	2394	1.84
MedDRA	23954	2.72	1739	2.33
ATC	5536	2.06	1588	1.11
MESH	29351	1.99	1460	1.69
ICD-O	3671	3.24	1122	2.88
DBPEDIA	912	1.78	381	1.85
ICPC	3046	7.09	235	2.26
ICF	3112	10.67	41	3.24
CLADIMED	4169	3.72	8	1.75
LOINC_2.66	91388	8.14	5	1.2
Total	499885	4.62	24242	3.35

Table 2: Reference terminologies and statistics regarding the validated sample of ICD-11. Number of sentences, average length in number of words (english corpus), number of included sentences in the validated sample of ICD-11, and their corresponding average length (number of words).

transferred to a new model. Specifically, the authors used an existing out-of-domain model trained on News data. Afterwards, they train their NMT system on the in-domain Biomedical'18 corpus[3].

Table 1 summarizes the related work on medical terms and texts translation, showing resources, family of machine translation approach, specific method used, languages studied and evaluation metrics, sorted by year.

3 Methodology

In the following section we describe the steps of our research methodology. First, a brief description of the terminologies and other corpora utilized is shown. Next, we describe the tools and libraries we have experimented with. Finally, the translation pipeline is presented.

3.1 Datasets

During our study we experimented upon numerous medical terminologies and datasets:

ATC (Anatomical Therapeutic Chemical, 2019). The ATC Classification System is a drug classification system that classifies the active ingredients of drugs according to the organ or system on which they act and their therapeutic, pharmacological and chemical properties. It is controlled by the World Health Organization Collaborating Centre for Drug Statistics Methodology (WHOCC), and was first published in 1976. Namely, the dataset includes

descriptions on metabolism, blood, dermatological and other contents.

CLADIMED (CLADIMED, 2019) is a five levels classification for medical devices, based on the ATC classification approach (same families). Devices are classified according to their main use and validated indications. It was originally developed by AP-HP (hospitals of Paris).

ACAD (Académie de Médecine, 2019). The "dictionnaire médical de l'académie de médecine" identifies terms used in health and defines them under the supervision of the French National Academy of Medicine.

ICD-O (World Health Organization, 2019). The International Classification of Diseases for Oncology (ICD-O) (1) has been used for nearly 35 years, principally in tumor or cancer registries, for coding the site (topography) and the histology (morphology) of the neoplasm, usually obtained from a pathology report.

MESH (Medical Subject Headings) (FR MESH, 2019) is a reference thesaurus in the biomedical field. The NLM (U.S. National Library of Medicine), which built it and updates it every year, uses it to index and query its databases, including MEDLINE/PubMed. INSERM, which has been the French partner of the NLM since 1969, translated the MeSH in 1986, and has been updating the French version every year since then. The bilingual version is often used as a translation tool, as well as for indexing and querying databases.

MedDRA (ICH, 2019) was developed in the late 1990s by the International Council for Harmonisation of Technical Requirements for Pharmaceuticals for Human Use (ICH). It constitutes a rich and highly specific standardised medical terminology to facilitate sharing of regulatory information internationally for medical products.

ORDO (Vasant et al., 2014). The Orphanet Rare Disease Ontology (ORDO) is a structured vocabulary for rare diseases derived from the Orphanet database, capturing relationships between diseases, genes and other relevant features. Orphanet was established in France by the INSERM (French National Institute for Health and Medical Research) in 1997. ORDO provides integrated, re-usable data for computational analysis.

[3]https://www.statmt.org/wmt18/biomedical-translation-task.html

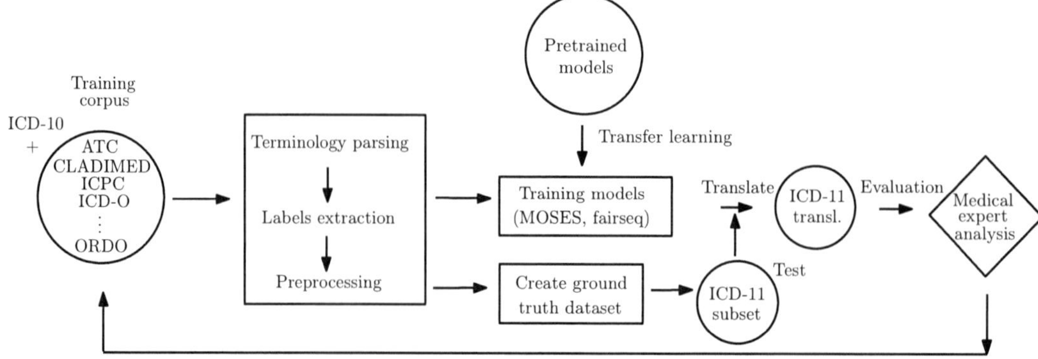

Figure 1: The proposed machine translation pipeline for ICD-11.

dbpedia (Auer et al., 2007). Through its API, dbpedia exposes multilingual fields and then can be used as a source to consolidate bi-lingual corpora.

ICD-10 (World Health Organization, 2016). ICD-10 is the 10th revision of the International Statistical Classification of Diseases and Related Health Problems (ICD), a medical classification list by the World Health Organization (WHO). It contains codes for diseases, signs and symptoms, abnormal findings, complaints, social circumstances, and external causes of injury or diseases. Work on ICD-10 began in 1983, endorsed by the Forty-third World Health Assembly in 1990, and it was first used by member states in 1994.

ICPC-2E (Verbeke et al., 2006). ICPC-2 classifies patient data and clinical activity in the domains of general/family medical practice and primary care, taking into account the frequency distribution of problems seen in these domains. It allows classification of the patient's reason for encounter, diagnostic, interventions, and the ordering of these data in an episode of care structure.

LOINC_2.66 (McDonald et al., 2003) is a widely used terminology standard for health measurements, observations, and documents.

CHU Rouen HeTOP is a large parallel corpus[4], including terminologies and ontologies in the domain of health, one of them being SNOMED CT[5].

ICF The International Classification of Functioning, Disability and Health (ICF), is a classification of health and health-related domains. ICF is the WHO framework for measuring health and disability at both individual and population levels.

In Table 2 we present the collection of medical terminologies and documents we explored during our research studies, as well as some statistics computed on them. We report size, average length of sentences in number of words, and number of sentences included in the validated sample of ICD-11.

3.2 Tools & libraries

Here we present publicly available tools that we used in our experiments. All the toolkits are written in Python, which offers a balance between complexity and usability. The Python community has increased dramatically during the past years, offering state-of-the-art methods in widely used libraries.

MOSES (Koehn et al., 2007) The MOSES tool software, is a phrasal-based probabilistic machine translation engine, which was used by many teams at the First Conference on Machine Translation (WMT16) (Bojar et al., 2016). Its base method includes word-alignment, phrase extraction and scoring during the training process.

fairseq (Ott et al., 2019) is a sequence modelling toolkit that allows researchers and developers to train custom models for translation, among other tasks. The toolkit offers a plethora of NMT models, like Long Short-Term Memory networks (LSTM) (Luong et al., 2015), Convolutional Neural Networks (CNN) (Dauphin et al., 2017; Gehring et al., 2017), as well as Transformer networks with self-attention (Vaswani et al., 2017; Ott et al., 2018).

Byte Pair Encoding (BPE) One of the most common problems in translating terminologies, including medical terminologies, are infrequent or unknown words, which the system has rarely or

---

[4]https://www.hetop.eu/hetop/
[5]http://www.snomed.org/

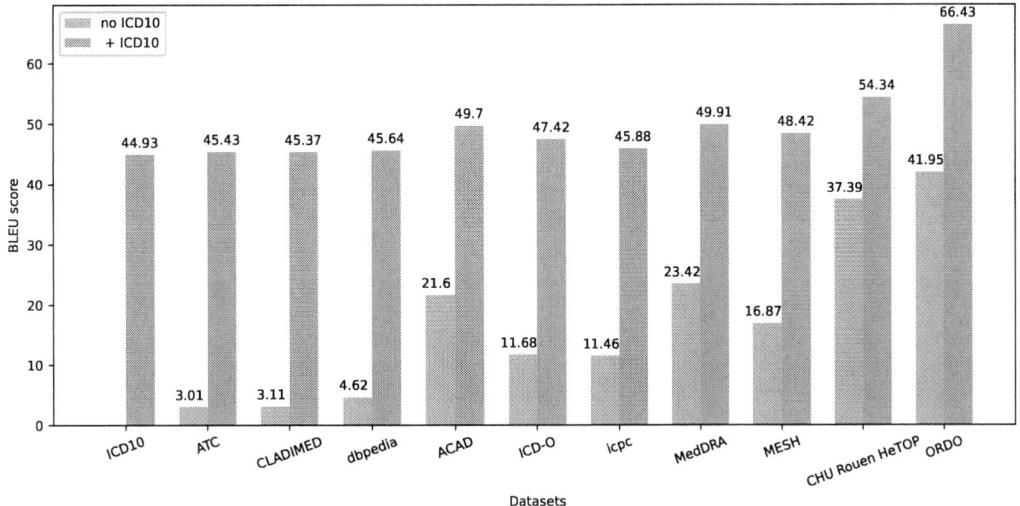

Figure 2: BLEU scores of MOSES on each dataset with and without ICD-10 in the training corpus using `multi-bleu.perl` by MOSES.

never seen. The effect is even more critical for NMT methods, where the vocabulary can not exceed the size of 50,000 or 100,000 words, due to the associated complexity. This limitation can be tackled by using subword units (BPE), a data compression technique (Sennrich et al., 2015). This step can be seen as part of preprocessing for the datasets, before training the models. We train our own BPE when no pre-trained model is used. In the transfer learning experiments, we use the provided BPE, as described in Section 4.

3.3 Dataset pipeline setup

An abstractive illustration of our proposed methodology is shown in Figure 1. Essentially, the pipeline can be split in five major parts: i) dataset & terminologies' search and retrieval, ii) parsing, extraction, preprocessing and extracting ground truth data, iii) model training, iv) translation and inspection, and v) evaluation and expert analysis.

Having access to the aforementioned datasets, we first applied terminology parsing. Next, we extracted the labels or descriptions, in order to form the corpus of parallel sentences. During the preprocessing step, we need to prepare the data for training the translation systems and perform tokenisation, truecasing and cleaning. For the NMT models, the BPE process is applied.

For ICD-11 given the fact that there is presently no human validated reference translation for French, we manually created one. The main ob-

Terminology	Size	avg_len(en)
ICD-11	123445	8.95
ICF	5920	10.79
Validated sample of ICD-11	24242	3.55

Table 3: Size (number of sentences) and average length in number of words (english corpus) for ICD-11, ICF and validated sample of ICD-11.

jective of our work is to examine how fast and effective a translation to a newly created or updated medical terminology can be developed, to be given to medical experts for preliminary evaluation work.

Our attempt offers the possibilities of speeding up the process of translating medical lexicons and documents, saving valuable human and computational resources. We evaluate our pipeline in two datasets: a sample of ICD-11 and the whole ICF terminologies. In the case of ICF terminology, we have access to both English and French medical experts validated versions. For ICD-11, since the French official version does not exist yet, we develop a method to evaluate and validate our results.

Through our studies, we discovered that a sample of the English ICD-11 terms can be found in existing French dictionaries. Thus, we can use these terms along with their French translation as already human-validated sentences. We end up having 24242 pairs in English and French that are already integrated in terminologies like ORDO, MESH_INSERM, LOINC_2.66 and others. Although, existing terms may as well require revi-

Method	Type	SacreBLEU ↑	BLEU ↑	METEOR ↑	TER ↓
MOSES no ICD10 (sys1)	SMT	39.92	35.61	33.84	50.61
MOSES only ICD10 (sys2)	SMT	45.84	39.16	35.18	45.22
MOSES dicts with ICD10 (sys3)	SMT	**65.59**	**57.50**	**46.20**	**28.62**
fairseq CNN no pre-trained (sys4)	NMT	51.02	42.93	38.85	38.98
fairseq CNN only pre-trained (sys5)	NMT	29.98	27.18	29.22	59.02
fairseq CNN finetuned on medical term/gies (sys6)	NMT	62.32	53.40	41.41	34.92
fairseq CNN finetuned on medical UFAL (sys7)	NMT	32.57	28.78	30.45	54.19

Table 4: SacreBLEU, BLEU, METEOR and TER scores on validated sample of ICD-11. Bold indicates best performance. SacreBLEU, BLEU and METEOR need to be maximized, while TER needs to be minimized.

Method	Type	SacreBLEU ↑	BLEU ↑	METEOR ↑	TER ↓
MOSES dicts with ICD10 (sys8)	SMT	50.40	42.82	38.74	38.50
fairseq finetuned on medical term/gies (sys9)	NMT	**60.82**	**52.46**	**42.97**	**32.59**

Table 5: Results on the ICD-11 24k sample, removed by the training dataset.

sion by a medical expert, the process indisputably accelerates the translation pipeline, compared to translating a terminology from scratch.

The automatic translation evaluation is based on the correspondence between the output and reference translation (ground truth/gold standard). We use popular metrics that cover several approaches:

- BLEU (Bilingual Evaluation Understudy) (Papineni et al., 2002) is calculated for individual translated segments (n-grams) by comparing them with a dataset of reference translations. Low BLEU score means high mismatch and higher score means a better match.

- SacreBLEU (Post, 2018) computes scores on detokenized outputs, using WMT (Conference on Machine Translation) tokenization and it produces the same values as the official script (`mteval-v13a.pl`) used by WMT.

- METEOR (Metric for Evaluation of Translation with Explicit ORdering) by Lavie and Agarwal (2007) includes exact word, stem and synonym matching while producing a good correlation with human judgement at the sentence or segment level (unlike BLEU which seeks correlation at the corpus level).

- TER (Translation Edit Rate) (Snover et al., 2006): the metric detects the number of edits (words deletion, addition and substitution) required to make a machine translation match exactly to the closest reference translation in fluency and semantics. High TER means high mismatch, while lower score means smaller distance from the reference text.

freq	sys3	sys6	len	sys3	sys6
1	0.8187	0.7858	-	-	-
2	0.8139	0.7626	<10	52.66	47.79
3	0.8263	0.7830	[10,20)	63.49	63.95
4	0.8429	0.7901	[20,30)	63.19	63.37
[5,10)	0.8521	0.8075	[30,40)	62.35	62.19
[10,100)	0.8714	0.8331	[40,50)	62.34	58.81
[100,1000)	0.7754	0.7749	[50,60)	59.64	59.82
≥1000	0.7773	0.7638	≥60	52.63	60.27

Table 6: Left: ICD-11 word accuracy analysis via `fmeasure` by frequency bucket. Right: sentence analysis by length bucket with BLEU metric for scoring.

Last, the translation is given to medical experts for analysis, recommending additional resources.

To the best of our knowledge, our work is one of the first that enables developing automatically a close to human-validated sample of a newly created or updated terminology. In Table 3 we present some statistics on our testing datasets.

4 Experiments & Results

In this section, we present the conducted experiments and obtained results. We selected two toolkits, due to their popularity and efficiency. MOSES represents the SMT tools, and fairseq represents the NMT domain. The summarized results of our experiments are visualized in Table 4. The traditional SMT model (sys3) manages to produce the best translation compared to the human validated sample, which consists mostly of short sentences. On the other hand, our best NMT model (sys6) performs slightly worse in total, but is better in longer sentences. The latter model (sys6) is finetuned on specialised medical terminologies, using as basis a largely pre-trained model on general do-

Method	Type	SacreBLEU ↑	BLEU ↑	METEOR ↑	TER ↓
MOSES dicts with ICD10 (sys3)	SMT	12.55	11.90	19.88	70.02
fairseq finetuned on medical term/gies (sys6)	NMT	**72.73**	**69.50**	**47.78**	**20.79**

Table 7: Results on translating the ICF terminology.

freq	sys3	sys6	len	sys3	sys6
1	0.3009	0.5323	-	-	-
2	0.2528	0.8251	<10	15.56	69.08
3	0.4284	0.8087	[10,20)	12.83	70.75
4	0.3315	0.8541	[20,30)	13.39	68.95
[5,10)	0.3501	0.8564	[30,40)	11.43	67.51
[10,100)	0.3812	0.8700	[40,50)	11.33	70.97
[100,1000)	0.5195	0.8761	[50,60)	6.44	69.93
≥1000	0.6644	0.8784	≥60	9.29	66.20

Table 8: Left: ICF word accuracy analysis via `fmeasure` by frequency bucket. Right: sentence analysis by length bucket with BLEU metric for scoring.

main corpora. In the next paragraphs we present our conducted experiments and results in detail.

MOSES We train our phrase-based translation system via MOSES, by building a 3-gram language model. First, we trained a model with all the medical terminologies excluding ICD-10 (sys1). We also experimented by using only ICD-10 (sys2) for training MOSES, reaching 44.93 in BLEU points. The model sys2 managed to perform better than any other dataset alone.

In order to identify the effectiveness of each terminology, we ran the translation process for each dataset separately, with and without ICD-10. Using only ATC, CLADIMED and dbpedia, resulted in poor performance, probably due to their specificity of included terms. Moreover, we observe that adding ICD-10 to all training datasets individually boosts the performance dramatically, as expected since many ICD-11 concepts come from ICD-10. Finally, training only on ORDO, we managed to reach a satisfying BLEU score. ORDO's effectiveness can be attributed to the large number of rare diseases it covers, which was one of the main improvements of ICD-11. The individual results are displayed in Figure 2.

Finally, we also trained an SMT model on the union of all the datasets. The model sys3 had the best performance, returning a high score of 65.59 SacreBLEU points, 57.50 BLEU points, 46.20 METEOR points and 28.62 TER points.

CNN trained on medical terminologies We trained a CNN model via fairseq on the medical terminologies we have gathered. The model (sys4)

reports a very good performance with 51.02 SacreBLEU points and 42.93 BLEU points. Nevertheless, as the number of training epochs was relatively small (30 epochs), the model may present an even better performance if trained for more epochs.

fairseq's pre-trained CNN model fairseq provides online pre-trained models for many language pairs, offering multiple architectures, trained on large amount of textual data[6].

For our experiments we selected the freely available `wmt14.en-fr.fconv-py` model (Gehring et al., 2017). The convolutional neural network (CNN) was trained on the WMT'14 English-French dataset. The full training set consisted of 35.5M sentence pairs, where sentences longer than 175 words were removed. Last, a size of 40K BPE types was selected for the source and target vocabulary. We used the same BPE types for encoding the test datasets in both languages. The model required 8 GPUs for about 37 days for training, as stated in Gehring et al. (2017).

The fairseq pre-trained model reports a low BLEU score, with 27.18 points, due to its general out-of-domain training data. Moreover, fairseq fails to translate all sentences in a satisfying manner. The phenomenon of extraneous translations, like "HAUT DE LA PAGE" or "PEPUDU", can be confirmed by searching analogous patterns across the output. To address this issue, we finetuned fairseq's CNN on medical terminologies.

fairseq's CNN finetuned on medical terminologies The finetuned model (sys6) incorporates transfer learning as it continues training the pre-trained CNN model by fairseq (Gehring et al., 2017), described in the previous paragraph, on medical terminologies, presented in Section 3.1. The model (sys6) almost reached the performance of the SMT approach, with a performance of 62.32 SacreBLEU points and 53.40 BLEU points, while being close to sys3 in both METEOR and TER points as well. As we will also present later in our analysis paragraph, the finetuned model (sys6) is better in translating long sentences (len>50) than

---

[6]`https://github.com/pytorch/fairseq/tree/master/examples/translation`

Ground truth/Reference	MOSES trained on medical terminologies (sys3)	fairseq CNN fine-tuned on medical terminologies (sys6)
pied convexe congénital bilatéral	pied convexe congénital bilatéral (100)	astragale verticale congénitale bilatérale (0)
syphilis des ostia coronaires	syphilis des ostia coronaires (100)	maladie ostiale coronarienne syphilitique (0)
chute accidentelle de la personne portée	personne portée (9.56)	chute accidentelle de la personne portée (100)
maladie des inclusions microvilleuses	atrophie microvillositaire congénitale (0)	maladie des inclusions microvilleuses (100)

Table 9: Translation examples of our trained models on the verified sample of ICD-11, given by `compare-mt`. The number in parenthesis shows the sentence translation score in BLEU points compared to reference.

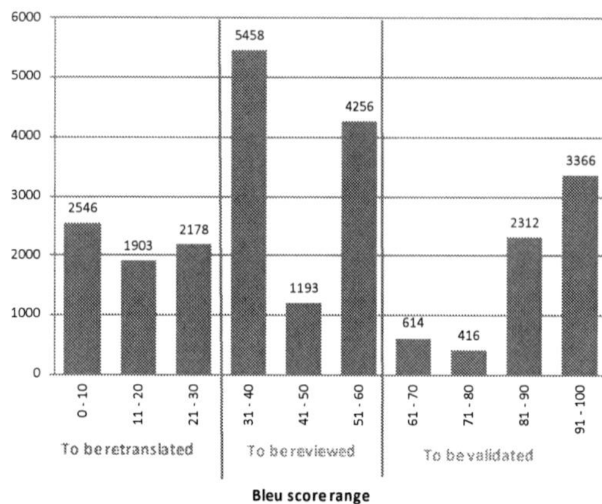

Figure 3: Sentence BLEU scoring on the 24k ICD-11 sample, categorized by a medical expert.

its MOSES rival (sys3), shown in Table 6. Our neural approach allows further training with no requirements.

fairseq's CNN finetuned on UFAL We also experimented on fine-tuning with the medical UFAL[7] dataset, a large medical domain corpus. The model (sys7) showed a performance of 28.78 BLEU points, being slightly better than using only the pre-trained CNN model. The low score can be attributed firstly to the short length nature of most ICD-11 sentences and secondly to the terminology syntax, which follows a specific structure. The medical UFAL consists mostly of long medical documents, which do not necessarily follow the typology of terminologies.

Removing the test sample from training As shown in Table 2, the validated sample of the ICD-11, which consists of 24k terms, is also included in the training dataset. Thus, we trained our two best architectures (sys3 & sys6) with removing the test set from the training corpus, creating two new models (sys8 & sys9). Table 5 presents their

performance, showing that the neural model is far superior from the statistical approach.

Testing on ICF Since the validated sample of ICD-11 was mostly known sentences of short size belonging to terminologies, we believe that the SMT approach will perform worse than NMT in generalizing to unknown terms and sentences. To confirm this hypothesis, we tested on ICF, where the average length is 10.79 and thus larger than the ICD-11 average length. We tested our two best models, MOSES trained with all the datasets (sys3) and the finetuned CNN fairseq model (sys6) toward the ICF terminology. The finetuned CNN (sys6) performs far better than MOSES (sys3), by a large difference, with 69.50 BLEU points compared to a low 11.90 BLEU points, respectively. sys6 is also far superior to sys3 in terms of METEOR and TER points. The scores are presented in Table 7.

Analysis We also analyzed our best methods with `compare-mt`[8] (Neubig et al., 2019) to study their output. The tool offers aggregate scoring with BLEU and other metrics, word accuracy via `fmeasure`[9], sentence bucket and n-gram difference analysis. Our analysis is summarized in Table 6. We see that the MOSES model (sys3) performance ranges depending the frequency of terms, while our finetuned CNN (sys6) remains stable, regardless of the frequency. Looking at the right part of Table 6, sys3 performs worse when the length of terms increases significantly (len>50), but remains better than its rival (sys6) for length<10.

Regarding the ICF terminology, results are shown in Table 8. We clearly observe that the finetuned CNN (sys6) manages to translate well all ICF terms regardless of their frequency on words. Moreover, looking at the right part of Table 8, while sys6 provides promising results with both short and long terms, sys3 (the MOSES model) struggles to perform well, especially when the length increases.

---

[7]`https://ufal.mff.cuni.cz/ufal_medical_corpus`

[8]`https://github.com/neulab/compare-mt`
[9]`https://en.wikipedia.org/wiki/F1_score`

BLEU score range	English label	fairseq proposal (sys6) **Human translations**	Comments
0-0,2	Familial hypophosphataemic rickets	rickets hypophosphatémiques familiaux **Rachitisme familial hypophosphatémique**	Unknown word
	Adult-onset Still disease, buttock	apparition d'un adulte maladie mortelle, fessier **Maladie de Still survenant chez l'adulte, fesse**	Proper name misunderstood/not recognised
	common bile duct blunt injury	lésion de contour du canal biliaire commun **blessure contondante du canal cholédoque**	ambiguity of label (common)
0,21-0,5	Context of assault, gang rivalry	contexte de l'agression, rivalité entre gangs **Contexte d'agression, rivalité entre gangs**	Inappropriate insertion of article
	Barrett adenocarcinoma	adénocarcinome barrett **Adénocarcinome de Barrett**	proper name misunderstood missing coordination term
	Fracture of thumb bone	fracture du pouce **Fracture de l'os du pouce**	missing word
0,51-0,9	ureter cyst	cyste de l'uretère **kyste de l'uretère**	unknown word translated with editorialy very close term
	talipes equinovalgus	pied bot equinovalgus **talipes equinovalgus**	use of correct synonym
	Unintentional exposure to or harmful effects of oxazolidinediones	exposition non intentionnelle ou effets nocifs des oxazolidinediones **Effets nocifs ou exposition accidentelle à des oxazolidinediones**	word order and use of correct synonym

Table 10: Comparison of translation outputs with human translations.

We also present translation examples coming from our trained models, based on `compare-mt`. Table 9 shows four examples of the translation systems. The first two lines present a perfect translation coming from MOSES (sys3), while the last two lines show a perfect translation by the finetuned CNN model (sys6), due to transfer learning.

Next, we present a categorization of the translation BLEU scores on the 24k ICD-11 validated sample in Figure 3. The translations were studied by a medical expert, who extracted three categories using manually selected thresholds. A relatively small 27% of the translations required re-translation, a 45% needs to be reviewed and finally a 28% require to be just validated.

Last, a comparison translation outputs with human translations follows in Table 10. We present translation examples, given by the finetuned CNN model with medical terminologies (sys6), compare them with human translations, observing interesting linguistic phenomena. The comparison shows that as the BLEU score increases, the system outputs "less acceptable" translations with cases like unknown words and ambiguities, to more "acceptable" translations with cases like word order and correct synonym use.

5   Conclusion

In this work, an automated pipeline for translating and evaluating medical terminologies is presented. The pipeline is tested comparing different machine translation methods, to translate WHO ICD-11 and ICF terminologies from English to French. Over ten legacy medical terminologies along with ICD-10 are used for training the pipeline. A traditional MOSES SMT approach that manages to produce a good baseline translation is shown. We have tested NMT methods and found that finetuning largely pre-trained models like fairseq's CNN on medical terminologies, incorporating transfer learning, can improve the quality of the translation. Last, we presented a simple method to generate automatically a test subset via existing terminologies.

The pipeline is adaptive to the typology of the studied terminology and it can be extrapolated easily to other languages for medical terminologies. The methodology enables researchers and healthcare end-users globally with a jump start approach that allows fast and effective translation of newly updated versions of terminologies.

For future work, using multilingual models (Liu et al., 2020) may omit the need for training multiple models in different languages. Last, additional medical datasets can be explored, not only for training but for creating larger validated corpora as well, following the constantly growing area of freely available language resources.

Acknowledgements

We would like to thank S. Darmoni from CHU Rouen / INSERM LIMICS (Paris) for sharing with us the HeTOP multilingual corpus that benefited the present study. We also thank the reviewers for their fruitful comments.

References

Académie de Médecine. 2019. Dictionnaire Médical de l'Académie de Médecine.

Anatomical Therapeutic Chemical. 2019. Atc.

Mihael Arcan and Paul Buitelaar. 2017. Translating domain-specific expressions in knowledge bases with neural machine translation. *arXiv preprint arXiv:1709.02184*.

Mihael Arcan, Mauro Dragoni, and Paul Buitelaar. 2016. Translating ontologies in real-world settings. In *International Semantic Web Conference*, pages 241–256. Springer.

Sören Auer, Christian Bizer, Georgi Kobilarov, Jens Lehmann, Richard Cyganiak, and Zachary Ives. 2007. Dbpedia: A nucleus for a web of open data.

Dzmitry Bahdanau, Kyunghyun Cho, and Yoshua Bengio. 2014. Neural machine translation by jointly learning to align and translate. *arXiv preprint arXiv:1409.0473*.

Ondřej Bojar, Rajen Chatterjee, Christian Federmann, Yvette Graham, Barry Haddow, Matthias Huck, Antonio Jimeno Yepes, Philipp Koehn, Varvara Logacheva, Christof Monz, et al. 2016. Findings of the 2016 conference on machine translation. In *Proceedings of the First Conference on Machine Translation: Volume 2, Shared Task Papers*, volume 2, pages 131–198.

CLADIMED. 2019. Classification des Dispositifs Médicaux (CLADIMED).

Vincent Claveau and Pierre Zweigenbaum. 2005. Translating biomedical terms by inferring transducers. In *Conference on Artificial Intelligence in Medicine in Europe*, pages 236–240. Springer.

Yann N Dauphin, Angela Fan, Michael Auli, and David Grangier. 2017. Language modeling with gated convolutional networks. In *Proceedings of the 34th International Conference on Machine Learning-Volume 70*, pages 933–941. JMLR.

Louise Deléger, Tayeb Merabti, Thierry Lecrocq, Michel Joubert, Pierre Zweigenbaum, and Stéfan Darmoni. 2010. A twofold strategy for translating a medical terminology into french. In *AMIA Annual Symposium Proceedings*, volume 2010, page 152. American Medical Informatics Association.

Louise Deléger, Magnus Merkel, and Pierre Zweigenbaum. 2009. Translating medical terminologies through word alignment in parallel text corpora. *Journal of Biomedical Informatics*, 42(4):692–701.

Ondřej Dušek, Jan Hajič, Jaroslava Hlaváčová, Michal Novák, Pavel Pecina, Rudolf Rosa, Aleš Tamchyna, Zdeňka Urešová, and Daniel Zeman. 2014. Machine translation of medical texts in the khresmoi project. In *Proceedings of the Ninth Workshop on Statistical Machine Translation*, pages 221–228.

Matthias Eck, Stephan Vogel, and Alex Waibel. 2004. Improving statistical machine translation in the medical domain using the unified medical language system. In *Proceedings of the 20th international conference on Computational Linguistics*, page 792. Association for Computational Linguistics.

FR MESH. 2019. Medical Subject Headings (MESH INSERM).

Jonas Gehring, Michael Auli, David Grangier, Denis Yarats, and Yann N Dauphin. 2017. Convolutional sequence to sequence learning. In *Proceedings of the 34th International Conference on Machine Learning-Volume 70*, pages 1243–1252. JMLR. org.

ICH. 2019. Medical Dictionary for Regular Activities.

Abdul Khan, Subhadarshi Panda, Jia Xu, and Lampros Flokas. 2018. Hunter nmt system for wmt18 biomedical translation task: Transfer learning in neural machine translation. In *Proceedings of the Third Conference on Machine Translation: Shared Task Papers*, pages 655–661.

Philipp Koehn, Hieu Hoang, Alexandra Birch, Chris Callison-Burch, Marcello Federico, Nicola Bertoldi, Brooke Cowan, Wade Shen, Christine Moran, Richard Zens, et al. 2007. Moses: Open source toolkit for statistical machine translation. In *Proceedings of the 45th annual meeting of the association for computational linguistics companion volume proceedings of the demo and poster sessions*, pages 177–180.

Audrey Laroche and Philippe Langlais. 2010. Revisiting context-based projection methods for term-translation spotting in comparable corpora. In *Proceedings of the 23rd international conference on computational linguistics*, pages 617–625. Association for Computational Linguistics.

Alon Lavie and Abhaya Agarwal. 2007. Meteor: An automatic metric for mt evaluation with high levels of correlation with human judgments. In *Proceedings of the second workshop on statistical machine translation*, pages 228–231.

Yinhan Liu, Jiatao Gu, Naman Goyal, Xian Li, Sergey Edunov, Marjan Ghazvininejad, Mike Lewis, and Luke Zettlemoyer. 2020. Multilingual denoising pre-training for neural machine translation. *arXiv preprint arXiv:2001.08210*.

Minh-Thang Luong, Hieu Pham, and Christopher D Manning. 2015. Effective approaches to attention-based neural machine translation. *EMNLP*.

Clement J McDonald, Stanley M Huff, Jeffrey G Suico, Gilbert Hill, Dennis Leavelle, Raymond Aller, Arden Forrey, Kathy Mercer, Georges DeMoor, John Hook, et al. 2003. Loinc, a universal standard for identifying laboratory observations: a 5-year update. *Clinical chemistry*, 49(4):624–633.

Graham Neubig, Zi-Yi Dou, Junjie Hu, Paul Michel, Danish Pruthi, Xinyi Wang, and John Wieting. 2019. compare-mt: A tool for holistic comparison of language generation systems. *CoRR*, abs/1903.07926.

Mikael Nyström, Magnus Merkel, Lars Ahrenberg, Pierre Zweigenbaum, Håkan Petersson, and Hans Åhlfeldt. 2006. Creating a medical english-swedish dictionary using interactive word alignment. *BMC medical informatics and decision making*, 6(1):35.

Jose Oncina. 1991. *Aprendizaje de lenguajes regulares y transducciones subsecuenciales*. Ph.D. thesis, PhD thesis, Universidad Politécnica de Valencia, Valencia, Spain.

Myle Ott, Sergey Edunov, Alexei Baevski, Angela Fan, Sam Gross, Nathan Ng, David Grangier, and Michael Auli. 2019. fairseq: A fast, extensible toolkit for sequence modeling. In *Proceedings of NAACL-HLT 2019: Demonstrations*.

Myle Ott, Sergey Edunov, David Grangier, and Michael Auli. 2018. Scaling neural machine translation. In *Proceedings of the Third Conference on Machine Translation (WMT)*.

Kishore Papineni, Salim Roukos, Todd Ward, and Wei-Jing Zhu. 2002. Bleu: a method for automatic evaluation of machine translation. In *Proceedings of the 40th annual meeting on association for computational linguistics*, pages 311–318. Association for Computational Linguistics.

Matt Post. 2018. A call for clarity in reporting BLEU scores. In *Proceedings of the Third Conference on Machine Translation: Research Papers*, pages 186–191, Belgium, Brussels. Association for Computational Linguistics.

Alejandro Renato, José Castano, Maria Avila Williams, Hernan Berinsky, Maria Gambarte, Hee Park, David Pérez, Carlos Otero, and Daniel Luna. 2018. A machine translation approach for medical terms. pages 369–378.

Rico Sennrich, Barry Haddow, and Alexandra Birch. 2015. Neural machine translation of rare words with subword units. *arXiv preprint arXiv:1508.07909*.

Mario J Silva, Tiago Chaves, and Barbara Simoes. 2015. An ontology-based approach for snomed ct translation. In *ICBO*.

Matthew Snover, Bonnie Dorr, Richard Schwartz, Linnea Micciulla, and John Makhoul. 2006. A study of translation edit rate with targeted human annotation. In *Proceedings of association for machine translation in the Americas*, volume 200. Cambridge, MA.

Ilya Sutskever, Oriol Vinyals, and Quoc V Le. 2014. Sequence to sequence learning with neural networks. In *Advances in neural information processing systems*, pages 3104–3112.

Drashtti Vasant, Laetitia Chanas, James Malone, Marc Hanauer, Annie Olry, Simon Jupp, Peter N Robinson, Helen Parkinson, and Ana Rath. 2014. Ordo: An ontology connecting rare disease, epidemiology and genetic data.

Ashish Vaswani, Noam Shazeer, Niki Parmar, Jakob Uszkoreit, Llion Jones, Aidan N Gomez, Łukasz Kaiser, and Illia Polosukhin. 2017. Attention is all you need. In *Advances in neural information processing systems*, pages 5998–6008.

Marc Verbeke, Diëgo Schrans, Sven Deroose, and Jan De Maeseneer. 2006. The international classification of primary care (icpc-2): an essential tool in the epr of the gp. *Studies in health technology and informatics*, 124:809.

Krzysztof Wołk and Krzysztof Marasek. 2015. Neural-based machine translation for medical text domain. based on european medicines agency leaflet texts. *Procedia Computer Science*, 64:2–9.

World Health Organization. 2016. ICD-10 : international statistical classification of diseases and related health problems : tenth revision.

World Health Organization. 2019. WHO ICD Oncology (ICDO).

Information retrieval for animal disease surveillance: a pattern-based approach

Sarah Valentin[1,2,3,4]**, Renaud Lancelot**[3,4] **and Mathieu Roche**[1,2]
[1]UMR TETIS, CIRAD, Montpellier, France.
[2]TETIS, AgroParisTech, CIRAD, CNRS, INRAE, Univ Montpellier, Montpellier, France.
[3]UMR ASTRE, CIRAD, Montpellier, France.
[4]ASTRE, CIRAD, INRAE, Univ Montpellier, Montpellier, France.
sarah.valentin28@gmail.com

Abstract

Animal diseases-related news articles are rich in information useful for risk assessment. In this paper, we explore a method to automatically retrieve sentence-level epidemiological information. Our method is an incremental approach to create and expand patterns at both lexical and syntactic levels. Expert knowledge input are used at different steps of the approach. Distributed vector representations (word embedding) were used to expand the patterns at the lexical level, thus alleviating manual curation. We showed that expert validation was crucial to improve the precision of automatically generated patterns.

1 Introduction

The ability to rapidly recognise emerging and re-emerging animal infectious diseases is a critical global health priority. Early warning is crucial for the quick implementation of effective control strategies at global and local levels (Heymann and Rodier, 2001). In recent decades, several outbreaks have highlighted the limitations of conventional disease surveillance, which is hampered by delayed detection and latency of the communication channels (Ben Jebara and Shimshony, 2006). The growing availability of digital data represents an unprecedented source of real-time disease information. Online news, social media and electronic health records are among the so-called informal sources that have proven to be valuable sources of disease information (Soto et al., 2008; Wilson and Brownstein, 2009). Their mainstreaming into surveillance systems, via the epidemic intelligence (EI) concept, has been a game-changer for disease surveillance and control. EI integrates two components in a single surveillance system: indicator-based surveillance (collection of structured data through traditional surveillance systems) and event-based surveillance (collection of unstructured data from informal sources, such as online news articles) (Paquet et al., 2006). Event-based surveillance (EBS) systems increasingly marshal text-mining methods to alleviate the amount of manual curation of the continuous flow of free text from informal sources (Hartley et al., 2010).

Animal-health online news articles are rich in different types of epidemiological information. For instance, news articles that report an outbreak often also describe outbreak control measures or economic impacts, share information about the outbreak source or draw attention to a given area at risk. Those elements may be of relevance to EI teams to assess risks associated with the occurrence of an outbreak.

In this paper, our objective is to value new types of epidemiological information contained in news. We describe a pattern-based method to automatically retrieve sentence-level epidemiological information. Empirically, sentence-level seems more homogeneous in terms of the topic than the whole news content. Our method is an incremental approach to create and expand patterns at both lexical and syntactic levels, with expert knowledge input. The learnt patterns can be further integrated into EBS systems.

2 Related work

2.1 Information retrieval from disease-related news

Information retrieval methods implemented in EBS systems broadly encompass three tasks: (1) document classification, (2) named entity extraction and (3) event extraction.

Most classification approaches in EBS systems focus on the binary news relevance, so as to filter out irrelevant ones (Conway et al., 2009; Doan et al., 2007; Torii et al., 2011; Valentin et al., 2020). Other classification frames assign a broad thematic

Proceedings of the 11th International Workshop on Health Text Mining and Information Analysis, pages 70–78
November 20, 2020. ©2020 Association for Computational Linguistics
https://doi.org/10.18653/v1/P17

label to the news, such as "outbreak-related" or "socioeconomic" (Zhang et al., 2009). When a news piece contains several topics, a single-label classifier has to decide on a topic (i.e. a label) among the other ones, which usually decreases the classification performance (Zhang et al., 2009).

Named entities extraction (NER) focuses on the detection of both epidemiological entities (e.g. virus, symptom) and domain unspecific entities (e.g. locations, dates). Event extraction consists in identifying the news narratives, by extracting the set of epidemiological entities describing a specific outbreak-event, also called attributes. The extraction of fine-grained temporal information such as the beginning and end of an event (Chanlekha et al., 2010), or thematic attributes such as the transmission mode (Conway et al., 2010).

None of the available tasks described here above suits the needs of our current work. Document-based classification is not precise enough to detect the variety of information contained in a single piece of news. Named entities and event extraction mainly focus on the spatio-temporal attributes of events, and partly address the potential of other types of epidemiological information. Midway between these two approaches, (Zhang and Liu, 2007) proposed a sentence-based annotation to detect outbreak-related sentences, while recognising that a news piece contains many sentences with different semantic meanings. However, as the primary goal was outbreak detection, outbreak-unrelated sentences (e.g. describing treatment or prevention) were all merged into one negative category. Based to this approach, we precisely aim to automatically retrieve epidemiological information at the sentence-level, broadening the binary outbreak-related/unrelated categories.

2.2 Pattern-based methods

In EBS systems, as well as in the biomedical literature, pattern-based approaches have been mainly used for named entities, event and relation extraction tasks (Wang et al., 2018). For instance, the EBS systems MedISys detects event from health-related news based on the Pattern-based Understanding and Learning System (PULS) (Du et al., 2016). PULS relies on a cascade of patterns applied to each sentence of the news article content to extract the event attributes. Patterns use both syntactical and semantic information of the sentence, such as:

NP(disease) VP(kill) NP(victim) ['in' NP(location)]

This pattern matches a noun phrase (NP) of semantic type, i.e. "disease"; a verb phrase (VP) headed by the verb "kill" (or its synonyms in the ontology) and has the adverbial phrase "so far", etc. The square brackets indicate an optional match. If the location is omitted in the sentence, it is inferred from the surrounding context. Verb phrases are not rigid and allow the presence of modifier elements, such as the auxiliary verb (e.g. "has") or adverb (e.g. "so far") (Steinberger et al., 2008).

Patterns can be generated manually or automatically. Manual construction typically relies on domain expert proposals, thereby achieving high precision. Such method is time-consuming and have two major shortcoming regarding the pattern generalisation: (i) the vocabulary is limited to the expert knowledge, and (ii) the syntactic structure may be rigid. Even if expert-based patterns may achieve high precision, the problem of recall, or coverage, is critical. Thus, weakly supervised and unsupervised methods have gained some popularity due to the marked reduction in the amount of manual curation they require (Ghosh et al., 2017; Ibekwe-Sanjuan et al., 2011; Yangarber, 2003). For instance, bootstrapping methods can generate patterns automatically from a pre-classified training corpus or from seed patterns (Jones et al., 1999; Thelen and Riloff, 2002). (Ibekwe-Sanjuan et al., 2011) used the local context of seed patterns, i.e. their surrounding words, to generate variants. This method relied on the assumption that patterns occurring in the same context tend to have the same semantic meaning, i.e. the paradigm of word embedding models. To our knowledge, the pattern-based approach was not evaluated to retrieve other types of epidemiological information in news articles. In line with (Ghosh et al., 2017; Ibekwe-Sanjuan et al., 2011), we propose an incremental approach integrating both word embedding and expert knowledge to expand patterns, in the sentence retrieval context. We propose two types of pattern expansion: (i) lexical and (ii) syntactical.

3 Method

3.1 Corpus

We used a publicly available corpus of news articles, that was annotated by four epidemiologists following specific guidelines (Valentin et al., 2019). The news articles are split in sentences and each sentence has two levels of annotation (i.e. two labels). The first annotation level aims at identifying the relevant sentences, i.e. the sentences describing a current, hypothetical (at risk) or past outbreak event. The second annotation level, called "Information type", characterises the epidemiological topic (fine-grained information). To evaluate retrieval methods on sufficient class sizes, we increased the annotated corpus (32 news articles, 486 sentences) with 56 additional news articles. We obtained a final corpus containing 1,245 sentences (from 87 news articles). From this initial corpus (1,245 sentences), 161 sentences were labelled as irrelevant. The subset of sentences for Information type classification hence consisted of 1,084 sentences (Table 1).

Table 1: Number of sentences per Information type category.

Category	No. of sentences
Protection and control measures	401
Descriptive epidemiology	310
Concern and risk factors	110
General epidemiology	109
Economic and political consequences	69
Transmission pathway	58
Distribution	27

3.2 Approach

In this section, we aim at identifying sentences from specific classes based on the patterns they contain. We opted for the pattern definition of Du and Yangarber, 2015, i.e. a pattern consists in "a place-holder for specific tokens (terms) and their surrounding context". The surrounding context may be fixed, and the token may be the variable. For instance, "**X** was detected in **Y**", where X is a disease and Y a location.

We evaluated a semi-automated and incremental process. In this approach, an expert is at the core of the process (Figure 1, steps 1 and 3). Indeed, we believe that expert knowledge is particularly suitable for the retrieval of fine-grained classes. Our objective is to use a minimal set of sentences (referred to as seed sentences) to identify patterns specific to the class (seed patterns) and expand the patterns at lexical and syntactical levels. All steps are detailed in the following subsections.

The pattern extracted and expanded after steps 1, 2, 3 and 4 are hereafter referred to as P_S (seed patterns), P_{E1} (first expansion), P_{E2} (second expansion), and P_{E3} (third expansion).

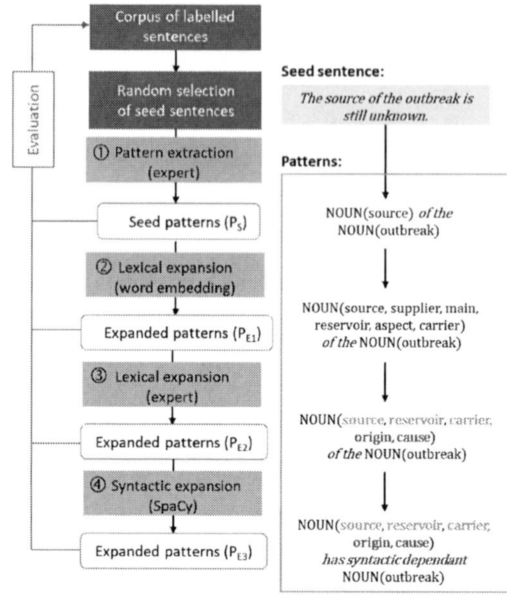

Figure 1: Incremental pattern expansion.
In the pattern box, seed terms are indicated between parentheses, and preceded by their POS. Terms validated by the expert are shown in green, and violet terms correspond to terms added by the expert. For readability, only the expansion of seed term 'source' is represented. The expansion steps (in orange) are detailed hereafter. Evaluation is done after each step.

Manual extraction of patterns (step 1) This first step aims at extracting an initial set of patterns based on a minimal subset of sentences (seed sentences). We relied on the expert to read each seed sentence and identified one or several patterns specific to the sentence class (seed patterns). For instance, based on the seed sentence from Figure 1, the identified pattern is:

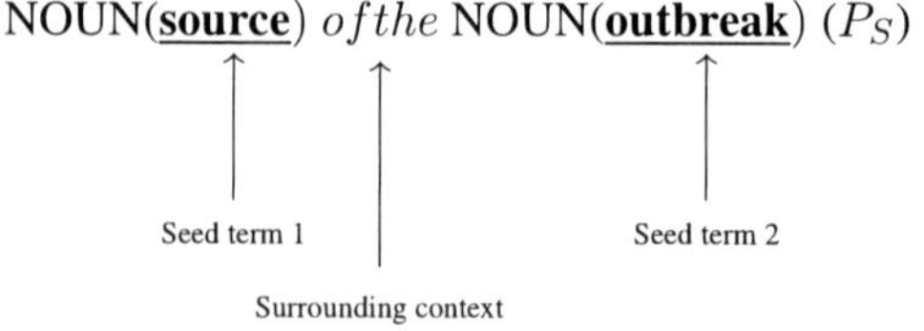

Where **source** and **outbreak** are the seed terms with *of* and *the* being linking words. Linking words usually consists of prepositions (e.g. "of", "through"), adverbs or auxiliary verbs. Seed terms include all terms present in the seed pattern. In our approach, seed terms include nouns, verbs (and their preposition), and adjectives.To match the patterns with sentences, we used the seed term lemmas labelled with their part-of-speech (POS) to avoid possible ambiguities between nouns and verbs. For instance, the seed term NOUN(cause) matches both nouns "cause" and "causes", but does not match conjugated forms of the verb "to cause", such as "causes" or "causes".

Several patterns appeared to involve disease or host as seed term, e.g. in "Investigators look for **swine fever links**". All disease names (including acronyms) and host were thus replaced in the text by the word "disease" and "host" respectively, so that they could be represented by NOUN(disease) and NOUN(host).

Automatic lexical expansion (step 2) At this step, we aim at expanding extracted patterns (P_s) at the lexical level. We focus on the seed terms, by automatically generating closed terms. We used a property of word embedding models, whereby words are represented as vectors. The vector values correspond to a densely distributed representation of the word. The values are learned according to the context in which the word appears, based on the assumption that words that frequently appear in the same context (i.e. surrounded by the same words) tend to have the same meaning (Goldberg, 2017). Words with common contexts have close vectors in the produced vector space. Several pre-trained word embeddings are publicly available, but training a word embedding model on corpus specific to the target domain has been shown to improve performances (Pyysalo et al., 2015). We thus decided to train a word embedding model on a dataset of animal-disease related news articles. We extracted the news from PADI-web database, published from February 20, 2014 to December 14, 2018. PADI-web is an open-source EBS system dedicated to animal health monitoring (Arsevska et al., 2018). We obtained a training set of 35,577 news articles. The training set length was 33,417,501 words, corresponding to a vocabulary of 464,536 single terms. The corpus was tokenized and lemmatized using the NLTK library (Bird and Loper, 2004). We trained the Word2Vec model, developed by Tomas Mikolov in 2013, which is one of the most popular techniques to learn word embedding (Mikolov et al., 2013). We used the continuous bag-of-words algorithm (CBOW) and we set the dimension for the trained vectors to 300, which is the dimension used in various studies (Mikolov et al., 2017, 2013; Pennington et al., 2014). We used the default parameter for the window size (5 words)— while setting the minimum word frequency at 10. In Word2Vec, the metric used to calculate the distance between two vectors is the standard cosine similarity (Leeuwenberg et al., 2016). The closer the cosine similarity between two word vectors is to one, the more similar the words are according to the model. Thus, for each word, cosine similarity can be used to rank terms in decreasing order of similarity.

For each seed term, we retrieved the K closest terms based on word embedding cosine similarity. For instance, the five closest terms for "source" are "supplier", "main", "reservoir", "aspect" and "carrier". The pattern is expanded based on the list of seed term synonyms. Finally, each **seed term** generates a set of K+1 variants (the seed term plus its close terms). Thus, if a pattern contains two seed terms generating K+1 variants each, we obtain $K+1^{K+1}$ combinations. We set K at 15, as a trade-off between the amount of input information and limited manual curation.

Manual lexical expansion (step 3) At the previous step, the K closest terms provided by the word embedding model were considered as seed term synonyms by default. At this step, we use expert knowledge to validate this assumption and enhance the lexical expansion at different steps:

1. Manual validation of the list of variants (K closest terms) generated automatically; terms judged as irrelevant or not specific enough regarding the sentence category are removed.

2. Adding a new term, if not present in the initial list of variants;

3. Merging seed terms (and their corresponding variants), when they are considered as synonyms.

The variants "supplier", "main", and "aspect" were considered as irrelevant by the expert and removed. The expert further added the variants "origin" and "cause" (Figure 1).

Structure expansion (step 4) The final step of manual curation consists of modifying the rigid contextual surrounding to improve the generalisation of patterns. A common approach is to use wildcards, i.e. symbols representing optional or specific characters and words in pattern matching. For instance, the pattern NOUN(source) of the NOUN(infection) could be replaced by:

NOUN(**source**) (W)? NOUN(**infection**)

The symbol (W)? indicates that NOUN(**source**) and NOUN(**infection**) can be separated by zero or more words. A major shortcoming of using wildcards is that syntactic information is not taken into account: the previous pattern only matches the source (or its variants) followed by the term infection (or its variants) but is not able to detect "the infection's source", for instance.

Thus, we proposed to expand the pattern structure based on the syntactic dependence between terms. More precisely, we modified the pattern structure when two (or more) seed terms were immediately syntactic dependent, i.e. connected by a single arc in the dependency tree (e.g. the subject of a verb, the adjective of a noun, etc.). In the previous example, the new pattern is:

NOUN(**source**) *has immediate syntactic dependant* NOUN(**infection**) *(P_{E3})*

This new pattern is now able to match both "the source of the infection" and "the infection's source". This approach has two advantages regarding pattern generalisation. First, as shown in the previous example, it increases the recall, as it does not rely on a specific sequence of terms such as wildcards. Second, the immediate syntactic relation a more fine-grained understanding of the meaning, thus avoiding irrelevant matches.

The analysis, including pattern-matching and generation of syntactic dependencies, was done using spaCy, i.e. a free open-source library for NLP in Python (Honnibal and Montani, 2017). We chose spaCy because it provides a pattern-matching function that readily allows pattern creation and enrichment. Syntactic relations were based on the spaCy dependency parser.

3.3 Evaluation

The pattern-based approach is particularly relevant for under-represented classes containing highly precise information, which may be hard to identify by supervised methods. In this context, we evaluated the pattern-based approach on two Information type classes, i.e. Concern and risk factors (CRF)

and Transmission pathway (TP). As seed sentences, we extracted 15% of an initial set of sentences belonging to the same class. The number of seed sentences was 16 (among 110) and 10 (among 69), respectively.

As shown in Figure 1, we evaluated the pattern performances after each expansion step in terms of recall, precision and F-measure. Here, as we focused on two categories, we evaluated the pattern approach as a binary classification—a sentence is classified as TP (resp. CRF) if it matches at least one TP (resp. CRF) patterns (positive sentence). Otherwise it is considered a negative sentence. We excluded seed sentences from the evaluation to avoid artificial positive matches. The testing dataset hence respectively consisted of 59 positive and 1,025 negative sentences for TP, and 94 positive and 990 negative sentences for CRF categories.

4 Results and discussion

4.1 Pattern-based retrieval

At step 1, the expert extracted 9 and 12 patterns from TP and CRF seed sentences, respectively (Table 2). No identifiable pattern could be found in two sentences (one in each category). In the CRF category, three sentences had the same pattern. After step 3 (manual lexical expansion), the number of terms represented 27% (65/240) and 68% (113/165) of the terms generated automatically in TP and CRF classes, respectively. For the same final number of patterns (7), the number of term variants in the CRF class was twofold higher in the CRF class than in the TP class.

Tables 3 and 4 show the performances of the pattern-based approach for the retrieval of TP and CRF sentences. The first extracted patterns (P_S) retrieved only 7% (4/59) of sentences and 47% (44/94) of CRF sentences.

In both classes, the precision decreased after the automatic lexical expansion (step 2) and increased after the manual expansion (step 3).

Manual and automatic pattern expansion did not impact the TP and CRF recall in similar ways. In the TP class, lexical expansions obtained mitigate improvement in recall, reaching a maximum value of 0.36 after step 3. The syntactic expansion increased the number of retrieved sentences by 179% (28 to 78 sentences), thus obtaining the highest recall of all steps (0.68).

In the CRF class, the automatic lexical expan-

Table 2: Numbers of patterns and terms at the different pattern expansion steps for both Transmission pathway (TP) and Concern and risk factor (CRF) categories. P_S, P_{E1}, P_{E2} and P_{E3} correspond to the seed patterns, and the 1st, 2nd and 3rd expansions, respectively.

	TP				CRF			
	P_S	P_{E1}	P_{E2}	P_{E3}	P_S	P_{E1}	P_{E2}	P_{E3}
No. of patterns	9	9	7	7	12	12	8	7
No. of terms (seeds and variants)	16	240	65	65	11	165	113	113

sion reached a recall of 0.80. Manual lexical and syntactic expansion did not improve the recall but contributed to increasing the precision from 0.33 to 0.53. The syntactic expansion did not impact the number sentences retrieved (+0.7%).

These results were consistent with the characteristics of the patterns extracted from both TP and CRF classes at the first step. In the CRF category, the seed terms mostly consisted of a noun such as threat, risk, or fear (14 out of 16 seed sentences). The recall stability among the lexical and syntactic expansion steps confirmed that CRF sentences were homogeneous regarding their syntactic structure and vocabulary. On the contrary, the extracted TP patterns were more complex. Seed terms mostly consisted of a verbal-linguistic structure such as "could have been brought by". Such syntactic structures were not generalizable, as highlighted by the poor recall at the first step.

4.2 Error analysis

Our results indicated that both lexical and syntactic features were crucial for improving the retrieval quality by the pattern-based approach. Yet the relative importance of each expansion step depended on semantic and syntactic specificity of each category. The manual curation by an expert allowed filtering out of irrelevant terms automatically generated by the word embedding model, thus improving the precision. However, the final precision (after step 4) did not exceed 0.53, indicating that some patterns were ambiguous and not sufficiently class-specific. This constraint was offset by the fact that we aimed to minimize the number of false-negative instances.

To understand what impacted the final recall, we manually evaluated the sentences not detected by the pattern-based approach (false negative sentences).

We found that 13/19 and 11/20 sentences were based on identifiable patterns that were not captured as seed patterns in step 1, in TP and CRF classes, respectively. For instance, four TP sentences referred to ongoing investigations about the outbreak's cause. Indeed, the term "investigators" was present in one of the seed sentences but was not identified by the expert as specific to the TP category.

In the second group of sentences (5/19 and 9/20 sentences), no specific patterns could be identified in, for instance:
"The minister said in one case a man bought meat from Ukraine and gave the water he washed the meat in to his pigs, which got sick and died."

This type of sentence underlines the limitations of the pattern-based approach. The classification decision is based only on matching with pre-defined terms (pattern seed terms), which are sometimes not sufficient to capture the whole semantics of the sentences. We hypothesise that supervised classification that takes all the sentence terms into account may perform better in such cases.

Eventually, one last TP sentence was not detected because it contained the pronoun "it" in reference to the disease, which did not match the list of expanded terms. This classic NLP problem is known as a noun phrase coreference resolution (Ng and Cardie, 2002). It highlights one limitation of the sentence-based approach in which references to entities are not inferred from other sentences. Further work could be focused on coreference resolution at the corpus level.

As a lexical expansion, we relied on terms automatically generated by the word embedding models. While a lot of retrieved terms were highly relevant, i.e. semantically close to the seed term, a substantial number of them were irrelevant, as shown by the drop in the number of variants after expert validation (step 3). A well-known alternative to retrieve term variants is the use of an external lex-

Table 3: Performances of the patterns for TP sentence retrieval in terms of precision, recall, and F-measure. The variation in the percentage corresponds to the change from the previous step.

	P_S	P_{E1}	P_{E2}	P_{E3}
Sentences retrieved (nb)	7	32 (+357%)	28 (-12,5%)	78 (+179%)
Precision	0.57 (4/7)	0.25 (10/32)	0.75 (21/28)	0.50 (39/78)
Recall	0.07 (4/59)	0.17 (10/59)	0.36 (21/59)	0.68 (40/59)
F-measure	0.12	0.20	0.49	0.58

Table 4: Performances of the patterns for CRF sentence retrieval in terms of precision, recall, and F-measure. The variation in the percentage corresponds to the change from the previous step.

	P_S	P_{E1}	P_{E2}	P_{E3}
Sentences retrieved (nb)	70	227 (+224%)	138 (-39%)	139 (+0.7%)
Precision	0.63 (44/70)	0.33 (75/227)	0.52 (72/138)	0.53 (74/139)
Recall	0.47 (44/94)	0.80 (75/94)	0.77 (72/94)	0.79 (74/94)
F-measure	0.53	0.47	0.62	0.63

ical database such as case WordNet (Luhn, 1958). However, as highlighted by Ibekwe-Sanjuan et al., 2011, WordNet is not domain-specific and may fail to provide appropriate words. For instance, the seed term "source" has "beginning", "root" and "informant" among its synonyms in WordNet. Word embedding models are also prone to generate irrelevant terms, i.e. not only retrieving synonyms but also antonyms, derived forms, hyper and hyponyms (Nooralahzadeh et al., 2018). Besides, there is no consensus regarding the number of terms to retrieve (K). Ghosh et al., 2017 used the top-5 terms to expand medical expression patterns. Relying on the same K for all seed terms unavoidably overlooks some relevant terms, or retrieves irrelevant ones, as the number of variants varies between terms. An alternative could be to set a minimum threshold for the cosine similarity value (Rekabsaz et al., 2017). But determining whether the similarity score obtained from word embedding is indicative of term synonymy is still an open question. Leeuwenberg et al., 2016 showed that cosine similarity alone is a bad indicator to determine if two words are synonymous. They proposed a new measure, i.e. relative cosine similarity, which calculates similarity relative to other cosine-similar words in the corpus.

5   Conclusion

In this paper, we presented an incremental approach to generate patterns for sentence-based retrieval, in the context of animal disease surveillance. The role of the expert was crucial to pinpoint the irrelevant terms generated by the word embedding model. Besides, the time cost of manual curation was minimal, as it mostly consisted of validating or adding terms. This time depends directly on the threshold chosen for top K retrieved terms. The use of syntactic dependency was easy to implement, thus alleviating the cost and bias of wildcard fine-tuning.

Contrary to the supervised approach, the pattern-based method is not hampered by so-called "black box" problems and can be easily enhanced by expert knowledge. Besides, even though it was not evaluated in this study, it is suitable for multi-label sentences as a sentence can match patterns from several classes. In future work, we aim at evaluating the pattern-based method on a new and larger corpus of news. Besides, it would be interesting to compare the pattern-based results with supervised sentence-based classifiers. We also think that the advantages of each method could be synergistic. A promising perspective could thus be to evaluate how to jointly take full advantage the strength of both methods. Cui et al., 2019 proposed an interesting approach to combine both supervised learning and manually built patterns and rules. They applied heuristic-based regular expression when the prediction confidence of the supervised classifier confidence was less than a specific threshold. Such an approach may help enhance the performance of Information retrieval in the event-based surveillance context.

6   Acknowledgments

This work was funded by the French General Directorate for Food (DGAL), the French Agricultural Research Centre for International Development (CIRAD) and the SONGES Project (FEDER and Occitanie). This work was supported by the French National Research Agency (ANR) under the Investments for the Future Program (ANR-16-CONV-0004).

References

Elena Arsevska, Sarah Valentin, Julien Rabatel, Jocelyn de Goër de Hervé, Sylvain Falala, Renaud Lancelot, and Mathieu Roche. 2018. Web monitoring of emerging animal infectious diseases integrated in the French Animal Health Epidemic Intelligence System. *PLOS ONE*, 13(8):e0199960.

M. Karim Ben Jebara and Arnon Shimshony. 2006. International monitoring and surveillance of animal diseases using official and unofficial sources. *Veterinaria Italiana*, 42(4):431–441.

Steven Bird and Edward Loper. 2004. NLTK: The Natural Language Toolkit. In *Proceedings of the ACL Interactive Poster and Demonstration Sessions*, pages 214–217, Barcelona, Spain. Association for Computational Linguistics.

Hutchatai Chanlekha, Ai Kawazoe, and Nigel Collier. 2010. A framework for enhancing spatial and temporal granularity in report-based health surveillance systems. *BMC medical informatics and decision making*, 10(1):1.

Mike Conway, Son Doan, Ai Kawazoe, and Nigel Collier. 2009. Classifying Disease Outbreak Reports Using N-grams and Semantic. *International Journal of Medical Informatics*, 78(12).

Mike Conway, Ai Kawazoe, Hutchatai Chanlekha, and Nigel Collier. 2010. Developing a Disease Outbreak Event Corpus. *Journal of Medical Internet Research*, 12(3):e43.

Menglin Cui, Ruibin Bai, Zheng Lu, Xiang Li, Uwe Aickelin, and Peiming Ge. 2019. Regular Expression Based Medical Text Classification Using Constructive Heuristic Approach. *IEEE Access*, 7:147892–147904.

Son Doan, Ai Kawazoe, and Nigel Collier. 2007. The Role of Roles in Classifying Annotated Biomedical Text. In *Biological, translational, and clinical language processing*, pages 17–24, Prague, Czech Republic. Association for Computational Linguistics.

Mian Du, Lidia Pivovarova, and Roman Yangarber. 2016. PULS: natural language processing for business intelligence. In *Proceedings of the 2016 Workshop on Human Language Technology and Intelligent Applications*, New York, United States.

Mian Du and Roman Yangarber. 2015. Acquisition of domain-specific patterns for single document summarization and information extraction. In *Proceedings of the The Second International Conference on Artificial Intelligence and Pattern Recognition*, Shenzhen, China.

Saurav Ghosh, Prithwish Chakraborty, Bryan L. Lewis, Maimuna S. Majumder, Emily Cohn, John S. Brownstein, Madhav V. Marathe, and Naren Ramakrishnan. 2017. Guided Deep List: Automating the Generation of Epidemiological Line Lists from Open Sources. *arXiv:1702.06663*.

Yoav Goldberg. 2017. Neural Network Methods for Natural Language Processing. *Synthesis Lectures on Human Language Technologies*, 10(1):1–309.

David Hartley, Noele Nelson, Ronald Walters, Ray Arthur, Roman Yangarber, Larry Madoff, Jens Linge, Abla Mawudeku, Nigel Collier, John Brownstein, Germain Thinus, and Nigel Lightfoot. 2010. The landscape of international event-based biosurveillance. *Emerging Health Threats Journal*, 3(0).

David L. Heymann and Guénaël R. Rodier. 2001. Hot spots in a wired world: WHO surveillance of emerging and re-emerging infectious diseases. *Lancet Infectious Diseases*, 1(5):345–353.

Matthew Honnibal and Ines Montani. 2017. spaCy 2: Natural language understanding with Bloom embeddings, convolutional neural networks and incremental parsing. In *To appear*.

Fidelia Ibekwe-Sanjuan, Fernandez Silvia, Sanjuan Eric, and Charton Eric. 2011. Annotation of Scientific Summaries for Information Retrieval. *arXiv:1110.5722*.

Rosie Jones, Andrew Mccallum, Kamal Nigam, and Ellen Riloff. 1999. Bootstrapping for Text Learning Tasks. In *In IJCAI-99 Workshop on Text Mining: Foundations, Techniques and Applications*, pages 52–63.

Artuur Leeuwenberg, Mihaela Vela, Jon Dehdari, and Josef van Genabith. 2016. A Minimally Supervised Approach for Synonym Extraction with Word Embeddings. *The Prague Bulletin of Mathematical Linguistics*, 105(1):111–142.

Hans Peter Luhn. 1958. The Automatic Creation of Literature Abstracts. *IBM Journal of Research and Development*, 2(2):159–165.

Tomas Mikolov, Edouard Grave, Piotr Bojanowski, Christian Puhrsch, and Armand Joulin. 2017. Advances in pre-training distributed word representations. *arXiv:1712.09405*.

Tomas Mikolov, Ilya Sutskever, Kai Chen, Greg Corrado, and Jeffrey Dean. 2013. Distributed Representations of Words and Phrases and Their Compositionality. In *Proceedings of the 26th International Conference on Neural Information Processing Systems - Volume 2*, NIPS'13, pages 3111–3119, USA. Curran Associates Inc.

Vincent Ng and Claire Cardie. 2002. Improving Machine Learning Approaches to Coreference Resolution. In *Proceedings of the 40th Annual Meeting of the Association for Computational Linguistics*, pages 104–111, Philadelphia, Pennsylvania, USA. Association for Computational Linguistics.

Farhad Nooralahzadeh, Lilja Øvrelid, and Jan Tore Lønning. 2018. Evaluation of domain-specific word embeddings using knowledge resources. In *Proceedings of the Eleventh International Conference on Language Resources and Evaluation (LREC 2018)*, Miyazaki, Japan. European Language Resources Association (ELRA).

Christophe Paquet, Daniel Coulombier, Reinhard Kaiser, and Marco Ciotti. 2006. Epidemic intelligence: a new framework for strengthening disease surveillance in Europe. *Eurosurveillance*, 11(12):5–6.

Jeffrey Pennington, Richard Socher, and Christopher Manning. 2014. Glove: Global Vectors for Word Representation. In *Proceedings of the 2014 Conference on Empirical Methods in Natural Language Processing (EMNLP)*, pages 1532–1543, Doha, Qatar. Association for Computational Linguistics.

Sampo Pyysalo, Tomoko Ohta, Rafal Rak, Andrew Rowley, Hong-Woo Chun, Sung-Jae Jung, Sung-Pil Choi, Jun'ichi Tsujii, and Sophia Ananiadou. 2015. Overview of the Cancer Genetics and Pathway Curation tasks of BioNLP Shared Task 2013. *BMC Bioinformatics*, 16(10):S2.

Navid Rekabsaz, Mihai Lupu, and Allan Hanbury. 2017. Exploration of a Threshold for Similarity Based on Uncertainty in Word Embedding. In *Advances in Information Retrieval*, volume 10193, pages 396–409. Springer International Publishing, Cham. Series Title: Lecture Notes in Computer Science.

Giselle Soto, Roger V. Araujo-Castillo, Joan Neyra, Miguel Fernandez, Carlos Leturia, Carmen C. Mundaca, and David L. Blazes. 2008. Challenges in the implementation of an electronic surveillance system in a resource-limited setting: Alerta, in Peru. In *BMC proceedings*, volume 2, page S4. BioMed Central.

Ralf Steinberger, Flavio Fuart, Erik van der Goot, Clive Best, Peter von Etter, and Roman Yangarber. 2008. Text Mining from the Web for Medical Intelligence. In *Mining Massive Data Sets for Security*. IOS Press.

Michael Thelen and Ellen Riloff. 2002. A bootstrapping method for learning semantic lexicons using extraction pattern contexts. In *Proceedings of the ACL-02 conference on Empirical methods in natural language processing - EMNLP '02*, volume 10, pages 214–221. Association for Computational Linguistics.

Manabu Torii, Lanlan Yin, Thang Nguyen, Chand T. Mazumdar, Hongfang Liu, David M. Hartley, and Noele P. Nelson. 2011. An exploratory study of a text classification framework for Internet-based surveillance of emerging epidemics. *International Journal of Medical Informatics*, 80(1):56–66.

Sarah Valentin, Elena Arsevska, Sylvain Falala, Jocelyn de Goër, Renaud Lancelot, Alizé Mercier, Julien Rabatel, and Mathieu Roche. 2020. PADI-web: A multilingual event-based surveillance system for monitoring animal infectious diseases. *Computers and Electronics in Agriculture*, 169:105163.

Sarah Valentin, Valérie De Waele, Aline Vilain, Elena Arsevska, Renaud Lancelot, and Mathieu Roche. 2019. Annotation of epidemiological information in animal disease-related news articles: guidelines and manually labelled corpus. *Dataverse Cirad*. Type: dataset.

Xuan Wang, Yu Zhang, Qi Li, Yinyin Chen, and Jiawei Han. 2018. Open Information Extraction with Meta-pattern Discovery in Biomedical Literature. In *Proceedings of the 2018 ACM International Conference on Bioinformatics, Computational Biology, and Health Informatics*, pages 291–300, Washington DC USA. ACM.

K. Wilson and J. S. Brownstein. 2009. Early detection of disease outbreaks using the Internet. *Canadian Medical Association Journal*, 180(8):829–831.

Roman Yangarber. 2003. Counter-training in discovery of semantic patterns. In *Proceedings of the 41st Annual Meeting on Association for Computational Linguistics*, volume 1, pages 343–350. Association for Computational Linguistics.

Yi Zhang and Bing Liu. 2007. Semantic text classification of emergent disease reports. In *Proceedings of the 11th European Conference on Principles and Pratice of Knockledge Discovery in Databases (PKDD)*, Warsaw, Poland. Springer.

Yulei Zhang, Yan Dang, Hsinchun Chen, Mark Thurmond, and Cathy Larson. 2009. Automatic online news monitoring and classification for syndromic surveillance. *Decision Support Systems*, 47(4):508–517.

Multitask Learning of Negation and Speculation using Transformers

Aditya Khandelwal
College of Engineering Pune
Pune, India
khandelwalar16.comp
@coep.ac.in

Benita Kathleen Britto
Veermata Jijabai Technological Institute
Mumbai, India
bcbritto_b16
@it.vjti.ac.in

Abstract

Detecting negation and speculation in language has been a task of considerable interest to the biomedical community, as it is a key component of Information Extraction systems from Biomedical documents. Prior work has individually addressed Negation Detection and Speculation Detection, and both have been addressed in the same way, using a 2 stage pipelined approach: Cue Detection followed by Scope Resolution. In this paper, we propose Multitask learning approaches over 2 sets of tasks: Negation Cue Detection & Speculation Cue Detection, and Negation Scope Resolution & Speculation Scope Resolution. We utilise transformer-based architectures like BERT, XLNet and RoBERTa as our core model architecture, and finetune these using the Multitask learning approaches. We show that this Multitask Learning approach outperforms the single task learning approach, and report new state-of-the-art results on Negation and Speculation Scope Resolution on the BioScope Corpus and the SFU Review Corpus.

1 Introduction

Detection of linguistic phenomena like Negation and Speculation are key components of Biomedical Information Retrieval systems, as they significantly alter the meaning of a sentence. While detecting these are also useful in Sentiment Analysis systems, and systems used to determine the veracity of information, their primary use is in biomedical systems. Thus, these tasks have attracted significant interest from researchers over the years, and due to the similarity between these tasks, similar approaches have been used to address them, and parallel corpora containing annotations for both Negation and Speculation have also been created, including:

- BioScope Corpus (Szarvas et al., 2008)

- SFU Review Corpus (Konstantinova et al.)

Prior research has converged to using a 2 stage approach for both Negation Detection and Speculation Detection: Cue Detection and Scope Resolution, and solved each task independently. These subtasks and their relevance to the Biomedical domain can be better understood using the following example:

(1) We found that T cells were [not] present, [perhaps] indicating an immuno deficiency.

Cue Detection involves finding the word(s) that express the linguistic phenomena being detected. In the example given above, *not* is the negation cue, as it expresses the negation in the sentence. Similarly, *perhaps* is the speculation cue.

Scope Resolution involves finding the word(s) that were affected by the cue word of the linguistic phenomena being considered. In the example above, the underlined words outline the scope for each cue. Specifically, for the negation cue *not*, the word *present* was negatively affected by it. Similarly, for the speculation cue *perhaps*, the words *indicating an immuno deficiency* were affected by it, indicating that these words have an associated uncertainty.

The approaches addressing these tasks have varied significantly over the years, with recent work focusing on using transformer-based architectures to perform transfer learning, and have given the best results to date on Scope Resolution. On Cue Detection, they yield the best performance among neural models, but due to the small dataset sizes, the best performance is still given by rule-based heuristic approaches.

Despite the similarity among the subtasks, prior systems have looked at these tasks independently. We believe that a system can improve performance on both tasks by learning from both simultaneously, due to the similarity, which is what Multitask Learning is about.

Multitask Learning involves jointly training the same architecture to perform multiple tasks. It

Proceedings of the 11th International Workshop on Health Text Mining and Information Analysis, pages 79–87
November 20, 2020. ©2020 Association for Computational Linguistics
https://doi.org/10.18653/v1/P17

relies on the concept of using shared knowledge between both tasks, which is what the model is forced to learn to perform well at all tasks, eventually leading to better performance on all tasks. For neural models, this is especially useful, as the lower layers can share the same input representation, thus getting more data to learn better lower level features from the input, and a task specific layer final layer can learn the task specific features.

In this paper, inspired by the success of transformers, we propose a method to perform Multitask Learning of negation and speculation using transformer-based architectures. We explore a few design choices, and analyse the impact of these design choices. We show that our approach provides significant benefits over the normal independently trained version. We also make all our code publicly available[1] . This paper is structured as follows: Section 2 contains a brief Literature Review, Section 3 describes the Methodology in detail, Section 4 talks about the Experimentation Details, Section 5 contains the Results and their Analysis, and Section 6 contains the Conclusion and Future Scope.

2   Literature Review

Over the years, methods addressing these subtasks have ranged from simple whitelists based on frequency (rule-based), to traditional Machine Learning algorithms like SVMs, neural models like BiLSTMS and transformer based models.

Khandelwal and Sawant (2020) provide an extensive literature review of the methods for Negation Cue Detection and Scope Resolution. For Speculation Cue Detection, most methods used were similar to the methods used for Negation Cue Detection. Below, we summarise a few papers that addressed Speculation Cue Detection and Scope Resolution.

2.1   Traditional Machine Learning Methods

Özgür and Radev (2009) used a Support Vector Machine (SVM) with a linear kernel to detect speculation cues. Once the speculation cues were identified, the parts-of-speech tags and the syntactic structure were used for scope resolution.

Morante and Daelemans (2009) used the IGTREE classifier with the help of gain ratio (TiMBL implementation) to classify the cues. For scope resolution, three classifiers were used to classify if a token was the first token in the scope sequence (F-scope), the last (L-scope), or neither. These three classifiers were Memory-based learning (as implemented in TiMBL), SVM and Conditional Random Field (CRF). A fourth classifier, the metalearner, used the output of the three classifiers to predict the scope classes.

Velldal et al. (2010) used a Maximum Entropy Classification approach to find the speculation cue. For scope resolution, they used a rule-based approach. The rules operated on the dependency structure of the parser (MaltParser and XLE).

Kilicoglu and Bergler (2008) used linguistic knowledge to detect speculation cues. This was achieved by using a semi-automatic lexical acquisition strategy as well as by using a dictionary of weighted speculation cues. In a follow-up paper (Kilicoglu and Bergler, 2010), they improved on their previous work with the help of vagueness quantifiers and syntactic dependency relations. Cues were detected with rules that operate on lexical information and syntactic information obtained from the Stanford Lexical Parser. The scopes of the cues were detected with the help of the Stanford Lexical Parser and dependency-based heuristics.

Velldal (2011) compiled a list of words that were observed to be cues in the training data, under the assumption that speculation cues can be treated as a closed class. He then checked occurrences of these words in the test data via a large-margin SVM classifier to determine whether they were a cue or not.

A CRF based approach was used in (Tang et al., 2010) to identify the hedge cues and their scopes in sentences. A CRF and large margin-based model were trained simultaneously. Their outputs were provided to another CRF model to get the cues of sentences. These cues were passed to another CRF model followed by post processing, to detect scopes of the given cues.

Morante et al. (2010) described a memory based approach (IGTree as implemented in TiMBL) for cue detection. For scope resolution, a memory-based approach was used with the help of syntactic dependencies of a sentence.

Read et al. (2011) described a methodology to resolve the scope of a sentence using an SVM based constituent ranker. The scope of a cue was assumed to be a constituent. Three broad classes of rules were used to extract features from the parse trees. The parse trees containing the cue were fed to the SVM-based ranker to output a ranked order of parse

---

trees which was then declared to be the scope of the cue.

Velldal et al. (2012) used an SVM classifier based on manually defined rules for speculation cue detection. For scope resolution, they experimented with three architectures: a rule-based system that used Data Driven Dependency Parsing to generate dependency structures, an SVM Ranker for selecting subtrees in the constituent structures obtained via a Grammar-Driven Phrase Structure Parser and hybrid of both the above systems.

The approach of (Moncecchi et al., 2012) was based on CRF and usage of domain knowledge. The task of cue detection was solved by using the sequential classifier of CRF. The scope of these cues was resolved by using the CRF at the initial stage with a window size of two. Later, domain knowledge was used to incorporate rules in the system, which showed an improvement in performance of the system.

Cruz et al. (2016) used a classifier-based approach for speculation cue detection and scope resolution. The features for the classifier were manually defined. For cue detection, an SVM-based classifier was used to predict the BIO tags. The scope was also identified using an SVM-based classifier to predict the in-scope and out-of-scope tags when the cues and tokens were provided as input to the classifier. A Radial Basis Function (RBF) Kernel was used with Cost Sensitive Learning to handle imbalanced classes.

2.2 Deep Learning Methods

Qian et al. (2016) used a CNN based approach to re-solve the scope of a speculation cue. The CNN framework took as input position and path features.

Fei et al. (2020) used a Recursive Neural Network (RecurNN) followed by a CRF to detect the scope in a sentence which is named as the Recur-CRF model. The dependency tree based RecurNN learnt a high-level representation of words in the given content. The output of the RecurNN was given to the CRF to fully under-stand the contextual information required to predict the scope of a given cue.

Recently, Britto and Khandelwal (2020) extended the approach by Khandelwal and Sawant (2020), who used BERT (Devlin et al., 2018) to address Negation Cue Detection and Scope Resolution. They experimented with using various transformer-based architectures (BERT, XLNet (Yang et al., 2019) and RoBERTa (Liu et al., 2019)), and jointly training on multiple datasets to address speculation cue detection and scope resolution. This approach gave the best results to date on Negation and Speculation Scope Resolution.

2.3 Multitask Learning using Negation Scope Resolution

We also review a couple of papers which have used Multitask Learning with Negation Scope Resolution as one of the many tasks to jointly train the model. It is important to note that these paradigms were explored to improve performance in the auxiliary tasks the model was trained, which were almost always harder than Negation Scope Resolution.

Bhatia et al. (2018) perform joint entity extraction and negation detection for biomedical articles. Initially, they used a hierarchical encoder-decoder model used for Named Entity Recognition (NER), and adapted it for the Multitask setting by sharing the encoder, but using separate decoders for both the tasks. To overcome the overparameterization during low-resource settings, they propose usage of a conditional softmax shared decoder, where instead of using 2 different decoder architectures, they shared the decoder as well, and only had separate classification heads. They also feed the output of the NER head as an additional input to the negation head, which helps improve the performance. They use BiLSTMs for both the encoder and decoder.

Barnes et al. (2019) explore another joint task that has been explored often, namely Sentiment Analysis systems that are jointly trained with Negation Detection systems. They mention that since Sentiment Analysis is a harder task than negation detection, and negation data is used as a task in the pipeline for sentiment analysis, they perform selective sharing of LSTM layers, and use negation as an auxillary task on which the sentiment analysis system is trained. Specifically, they use a separate CRF tagger for negation detection on the outputs of an intermediate layer for the sentiment analysis system, whose final layer is used for sentiment classification. They use a BiLSTM-based network.

3 Methodology

Similar to (Khandelwal and Sawant, 2020) and (Britto and Khandelwal, 2020), we use the transformer model (BERT/XLNet/RoBERTa) with a classification head as our base model. To jointly train the model, we propose the following additions to the model.

3.1 Cue Detection

For Cue Detection, we use 2 separate classification heads for Negation Cue Detection and Speculation Cue Detection respectively. The architecture can be visualized as in Figure 1. We feed an input

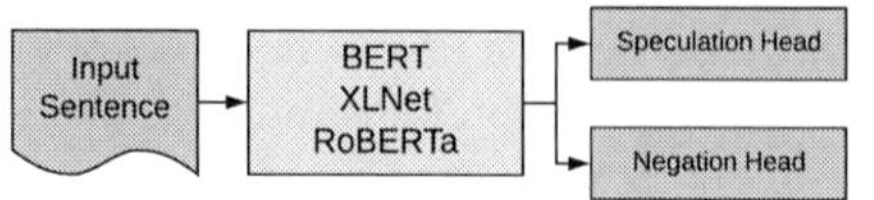

Figure 1: Multitask Cue Detection (Model Overview)

sentence to the model, and use the output corresponding to the task we are looking to perform. This architecture halves the number of parameters and inference time if we want to perform negation and speculation detection simultaneously.

To train this model, we only train on those sentences that have both negation and speculation cue labels. Since we train on the BioScope Corpus, and the SFU Review Corpus, all training samples have labels for both negation and speculation. A single input sentence is fed, and the model is trained on the losses computed for both heads, negation and speculation.

3.2 Scope Resolution

For Scope Resolution, we use the same classification head for both Negation and Speculation Scope Resolution, and use preprocessing techniques to implicitly tell the model which task to perform.

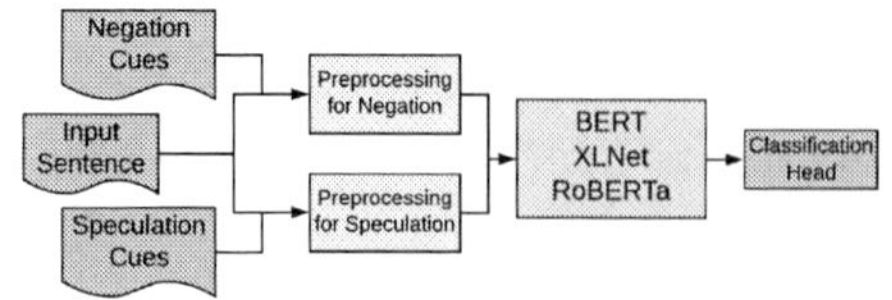

Figure 2: Multitask Scope Resolution (Model Overview)

For Scope Resolution, we need to represent the cue words in the input sentence for which we want to find the scope. This could be done via the Augment Preprocessing method used by (Britto and Khandelwal, 2020). This involves appending a special token before the cue word in the input sentence

which represents the type of cue word. The types of cue words considered are:

- Single Word Cue: tok[0]

- Part of a Multiword Cue: tok[1]

- Affix (Suffix / Prefix): tok[2]

Consider the following example:

Input Sentence: *It seems that the treatment is not successful.*
Negation Cues: *not*
Preprocessed Sentence: *It seems that the treatment is tok[0] not successful.*

To jointly train the same model to make predictions, we have to also tell the model which task we expect it to perform. To do this, we propose the following methods which are slight modifications of the Augment preprocessing method:

- Global: We represent the task by appending the name of the task to be performed at the end of the input sentence followed by a [SEP] token. The cue words for both negation and speculation are represented by the same set of special tokens. Specifically,

  Input Sentence: *It seems that the treatment is not successful.*
  Negation Cues: *not*
  Speculation Cues: *seems*
  Input Sentence for Negation: *It seems that the treatment is tok[0] not successful [SEP] Negation.*
  Input Sentence for Speculation: *It tok[0] seems that the treatment is not successful [SEP] Speculation.*

  Thus, the type of cue for both negation and speculation is the same (single word cue), hence we use the same token (tok[0]) to augment the input sentence. The task is represented by appending the task name to the end of the sentence.

- Local: Here, we use the following tokens to represent the different types of negation and speculation cues.

  – Negation-Single Word Cue: tok[0]

- Negation-Part of a Multiword Cue: tok[1]
- Negation-Affix (Suffix / Prefix): tok[2]
- Speculation-Single Word Cue: tok[4]
- Speculation-Part of a Multiword Cue: tok[5]
- Speculation-Affix (Suffix / Prefix): tok[6]

Specifically,

Input Sentence: *It seems that the treatment is not successful.*
Negation Cues: *not*
Speculation Cues: *seems*
Input Sentence for Negation: *It seems that the treatment is tok[0] not successful.*
Input Sentence for Speculation: *It tok[4] seems that the treatment is not successful.*

Here, the tokens used to represent different types of negation cues are different than the tokens used to represent the different types of speculation cues, thus implicitly telling the model which scope it has to find.

4 Experimentation Details

We perform experimentation on the following datasets:

- BioScope Corpus:
    - BioScope Abstracts (BA) SubCorpora
    - BioScope Full Papers (BF) SubCorpora
- SFU Review Corpus (SFU)

We believe that by training on multiple datasets, the overfitting of the models can reduce, as the datasets are fairly small in size (200-2000 samples), despite the different domains of the datasets (BioScope Corpora is from the Biomedical Domain, and SFU Review Corpus contains general online review text). Hence, we also experiment with training the models on multiple datasets, and testing on the individual datasets.

We use a 70-15-15 train-dev-test split. The results are reported as an average of 5 runs for training on a single dataset and an average of 3 runs for training on a combination of multiple datasets. We report the Macro F1 Average (Token-level) score for both Cue Detection and Scope Resolution.

We perform an early stopping (with a patience of 6) on the validation F1 Score. Since we jointly train 2 tasks, we experiment with these 2 ways to perform early stopping:

- Separate: Here, we use 2 early stopping counters: One for Negation and one for Speculation. Specifically, we have separate validation sets for Negation and Speculation, and for each validation set, we run a different Early Stopping Counter. Thus, the final models for Negation and Speculation differ, although they are trained jointly.

- Combined: Here, there is only one Early Stopping used. Training is stopped when the average of the validation F1 scores on the Negation Validation set and the Speculation Validation set do not improve for 6 epochs. Here, we have the same final model for both Negation and Speculation.

We train the models using GPUs available via Google Colaboratory. The code is publicly available.

5 Results and Analysis

To perform a better comparison of independently trained models on multiple datasets, we train BERT, XLNet and RoBERTa on BF+BA, BF+SFU, BA+SFU and BF+BA+SFU for Negation Cue Detection and Negation Scope Resolution. We train the model as per the paper by Britto and Khandelwal (2020), and average the results of 3 runs. The results are shown in Tables 1 and 2. An analysis of the results shown below is done in Section 5.4.

Test Dataset	Model	Train Dataset			
		BF+BA	BF+SFU	BA+SFU	BF+BA+SFU
BA	BERT	93.27	93.90	93.20	**89.92**
BA	RoBERTa	92.42	93.58	92.86	88.01
BA	XLNet	**95.04**	**96.42**	**94.74**	89.85
BF	BERT	88.74	91.05	87.94	**91.99**
BF	RoBERTa	87.66	92.66	89.93	86.87
BF	XLNet	**89.33**	**94.60**	92.83	88.17
SFU	BERT	**85.74**	**50.89**	**84.72**	84.05
SFU	RoBERTa	83.74	17.70	82.89	71.98
SFU	XLNet	77.72	31.96	73.07	**86.01**
Negation Cue Detection					

Table 1: Negation Cue Detection (Trained on Multiple Datasets)

5.1 Cue Detection

The results for Negation and Speculation Cue Detection are shown in Table 3 (trained using the Com-

Test Dataset	Model	Train Dataset			
		BF+BA	BF+SFU	BA+SFU	BF+BA+SFU
BA	BERT	94.24	88.22	94.45	90.17
	RoBERTa	94.67	92.84	94.11	90.54
	XLNet	**94.84**	**96.77**	**96.03**	**92.58**
BF	BERT	90.01	81.91	90.74	87.91
	RoBERTa	90.76	91.51	94.63	86.84
	XLNet	**92.18**	**95.73**	**97.12**	**92.07**
SFU	BERT	90.19	89.96	**85.98**	89.71
	RoBERTa	90.08	**90.83**	85.60	**91.34**
	XLNet	**90.74**	89.83	85.89	89.70
Negation Scope Resolution					

Table 2: Negation Scope Resolution (Trained on Multiple Datasets)

Test Dataset	Model	Train Dataset						
		BA	BA+SFU	BF	BF+BA	BF+BA+SFU	BF+SFU	SFU
BA	BERT	88.73	87.80	87.83	91.13	90.00	54.75	66.84
	RoBERTa	93.06	91.90	91.35	93.91	93.44	89.50	82.56
	XLNet	**95.70**	**94.53**	**92.92**	**97.01**	**96.08**	**91.92**	**84.08**
BF	BERT	83.90	82.65	84.42	86.87	81.60	75.57	75.11
	RoBERTa	88.21	87.28	89.79	91.24	91.65	86.69	79.52
	XLNet	**91.71**	**89.98**	**90.89**	**96.25**	**94.30**	**88.76**	**79.78**
SFU	BERT	24.67	80.55	30.27	29.82	72.75	77.35	58.08
	RoBERTa	23.21	85.09	23.09	32.19	**83.38**	83.90	78.98
	XLNet	**32.22**	**86.88**	**30.70**	**35.45**	73.38	**86.35**	**86.21**
Negation Cue Detection								

Test Dataset	Model	Train Dataset						
		BA	BA+SFU	BF	BF+BA	BF+BA+SFU	BF+SFU	SFU
BA	BERT	86.91	81.84	78.56	86.40	84.09	62.42	53.11
	RoBERTa	90.36	89.48	86.48	90.57	89.17	83.52	57.71
	XLNet	**93.66**	**92.75**	**90.29**	**93.98**	**92.94**	**89.59**	**58.50**
BF	BERT	72.18	65.38	71.16	73.76	65.77	58.31	51.97
	RoBERTa	78.53	73.03	80.29	84.25	79.63	78.11	55.04
	XLNet	**84.00**	**81.06**	**84.13**	**87.00**	**87.14**	**82.01**	**58.39**
SFU	BERT	21.27	90.39	23.81	23.90	77.78	89.25	90.76
	RoBERTa	23.82	86.72	22.60	30.23	**88.39**	88.42	87.90
	XLNet	**29.34**	**92.36**	**27.57**	**33.04**	74.12	**92.10**	**92.61**
Speculation Cue Detection								

Table 3: Results for Cue Detection (Combined Early Stopping)

bined Early Stopping method), and Table 4 (trained using the Separate Early Stopping method). We compare the Combined and Early Stopping methods in Section 5.4.

Test Dataset	Model	Train Dataset						
		BA	BA+SFU	BF	BF+BA	BF+BA+SFU	BF+SFU	SFU
BA	BERT	91.10	91.00	87.91	91.90	90.17	75.27	68.73
	RoBERTa	94.21	93.15	90.44	93.30	93.02	88.08	82.45
	XLNet	**95.98**	**95.57**	**93.23**	**96.27**	**94.86**	**91.85**	**84.51**
BF	BERT	85.34	85.15	86.89	88.55	85.03	74.60	76.13
	RoBERTa	90.61	88.45	89.16	90.42	89.18	89.76	79.66
	XLNet	**92.14**	**90.37**	**91.89**	**92.57**	**94.37**	**90.18**	**79.66**
SFU	BERT	**38.61**	**82.96**	**54.17**	26.40	84.42	83.82	77.57
	RoBERTa	33.37	83.82	11.62	11.34	61.98	83.79	**82.42**
	XLNet	19.16	61.09	31.00	**27.28**	**87.07**	**86.14**	64.29
Negation Cue Detection								

Test Dataset	Model	Train Dataset						
		BA	BA+SFU	BF	BF+BA	BF+BA+SFU	BF+SFU	SFU
BA	BERT	84.84	79.72	79.49	88.23	83.66	62.04	52.41
	RoBERTa	90.81	88.80	86.95	89.00	86.95	85.55	57.99
	XLNet	**93.11**	**92.58**	**89.39**	**93.17**	**91.48**	**88.74**	**61.41**
BF	BERT	72.42	65.38	74.00	78.21	69.66	64.84	51.13
	RoBERTa	79.17	73.97	82.96	80.42	76.63	79.93	55.30
	XLNet	**83.10**	**81.34**	**86.94**	85.78	**88.63**	**82.46**	**59.64**
SFU	BERT	29.67	**89.71**	**41.43**	21.38	89.01	90.67	83.02
	RoBERTa	**33.28**	88.33	13.26	12.74	58.26	88.05	**86.36**
	XLNet	18.74	62.16	26.96	**25.61**	**92.14**	**92.41**	63.50
Speculation Cue Detection								

Table 4: Results for Cue Detection (Separate Early Stopping)

A comparison of the best models trained jointly on Negation and Speculation compared with the independently trained model variants and the state-of-the-art results is shown in Table 5.

Task	Dataset	Model	Author	F1
Negation Cue Detection	BioScope Abstracts	ML Classifier	Morante, Daelemans	**98.68**
		XLNet (BF+SFU) (Independently Trained)	-	96.42
		XLNet (BF+BA) (Jointly Trained)	Ours	97.01
	BioScope Full Papers	ML Classifier	Morante, Daelemans	**97.81**
		XLNet (BF+SFU) (Independently Trained)	-	94.60
		XLNet (BF+BA) (Jointly Trained)	Ours	96.25
	SFU	ML Classifier	Cruz, Taboada, Mitkov	**89.64**
		XLNet (SFU) (Independently Trained)	Britto, Khandelwal	87.32
		XLNet (BF+BA+SFU) (Jointly Trained)	Ours	87.07
Speculation Cue Detection	BioScope Abstracts	SVM	Ozgur, Radev	91.69
		XLNet (BF+BA) (Independently Trained)	Britto, Khandelwal	**95.61**
		XLNet (BF+BA) (Jointly Trained)	Ours	93.98
	BioScope Full Papers	SVM	Ozgur, Radev	82.82
		XLNet (BF+BA+SFU) (Independently Trained)	Britto, Khandelwal	**93.84**
		XLNet (BF+BA+SFU) (Jointly Trained)	Ours	88.63
	SFU	SVM	Diaz, Taboada, Mitkov	92.37
		BERT (SFU) (Independently Trained)	Britto, Khandelwal	**92.66**
		XLNet (BF+SFU) (Jointly Trained)	Ours	92.61

Table 5: Comparison of Cue Detection Results with State-of-the-Art Results

5.2 Negation Scope Resolution

The results for Negation Scope Resolution are shown in Table 6 (trained using the Combined Early Stopping method) and Table 7 (trained using the Separate Early Stopping method). We compare the Combined and Early Stopping methods in Section 5.4.

A comparison of the best models trained jointly on Negation and Speculation compared with the state-of-the-art results for Negation Scope Resolution is shown in Table 8. Our joint training approach outperforms the existing state-of-the-art models (independently trained transformer based architectures) on all datasets that we experiment with.

5.3 Speculation Scope Resolution

The results for Speculation Scope Resolution are shown in Table 9 (trained using the Combined Early Stopping method) and Table 10 (trained using the Separate Early Stopping method). We compare the Combined and Early Stopping methods in Section 5.4.

Test Dataset	Model	Train Dataset						
		BA	BA+SFU	BF	BF+BA	BF+BA+SFU	BF+SFU	SFU
BA	BERT	94.33	95.48	92.25	95.67	94.91	91.02	84.40
	RoBERTa	95.08	93.80	92.08	93.41	93.77	89.82	83.17
	XLNet	**96.68**	**96.21**	**94.42**	96.19	**97.06**	**91.51**	84.11
BF	BERT	91.41	90.36	91.10	**97.40**	93.00	88.57	79.94
	RoBERTa	92.74	89.73	90.72	96.53	92.43	89.74	78.28
	XLNet	**95.43**	**93.11**	**93.67**	96.92	95.17	**93.03**	**80.18**
SFU	BERT	85.47	90.62	**85.18**	**85.77**	92.07	91.45	91.34
	RoBERTa	85.05	**92.37**	84.20	84.58	**93.19**	90.19	91.31
	XLNet	**86.16**	91.37	84.11	85.63	91.19	92.69	**91.41**
Negation Scope Resolution: Global								

Test Dataset	Model	Train Dataset						
		BA	BA+SFU	BF	BF+BA	BF+BA+SFU	BF+SFU	SFU
BA	BERT	94.46	94.13	91.52	94.18	95.29	90.86	83.30
	RoBERTa	94.21	93.78	91.59	94.64	93.43	90.84	83.27
	XLNet	**95.89**	**95.89**	**94.82**	96.30	96.01	**93.59**	**83.94**
BF	BERT	92.61	90.38	92.85	94.44	94.02	89.32	79.72
	RoBERTa	92.64	91.28	91.28	96.28	94.28	89.63	78.90
	XLNet	**94.62**	**93.03**	**94.81**	96.78	96.09	**93.11**	**80.07**
SFU	BERT	85.91	**91.57**	**85.31**	**85.98**	91.15	91.39	90.84
	RoBERTa	84.85	91.22	83.85	84.68	91.03	90.27	**91.71**
	XLNet	85.21	91.37	83.99	85.17	**91.49**	**91.98**	91.51
Negation Scope Resolution: Local								

Table 6: Results for Negation Scope Resolution (Combined Early Stopping)

Test Dataset	Model	Train Dataset						
		BA	BA+SFU	BF	BF+BA	BF+BA+SFU	BF+SFU	SFU
BA	BERT	94.27	92.79	92.00	94.50	93.55	90.83	83.75
	RoBERTa	94.55	93.23	91.45	94.46	94.77	89.56	83.05
	XLNet	**96.15**	**96.62**	**94.52**	96.97	95.92	**94.10**	**83.97**
BF	BERT	91.93	89.18	90.66	94.46	93.39	86.61	79.65
	RoBERTa	92.25	90.04	91.89	95.93	92.02	90.16	78.83
	XLNet	**94.47**	**92.91**	**95.40**	96.55	96.62	**91.65**	**80.16**
SFU	BERT	85.28	**91.90**	84.93	85.84	91.19	**91.90**	**91.93**
	RoBERTa	85.01	91.19	83.49	85.42	**91.29**	90.12	91.19
	XLNet	**85.38**	90.45	84.69	85.40	90.38	91.70	90.98
Negation Scope Resolution: Global								

Test Dataset	Model	Train Dataset						
		BA	BA+SFU	BF	BF+BA	BF+BA+SFU	BF+SFU	SFU
BA	BERT	94.73	94.17	91.46	94.74	94.19	91.11	**83.49**
	RoBERTa	94.21	93.89	91.26	95.61	94.59	90.27	83.27
	XLNet	**96.22**	**96.80**	**94.30**	96.08	96.33	**93.14**	83.00
BF	BERT	92.55	90.85	**93.52**	96.16	91.62	89.97	79.55
	RoBERTa	92.64	90.19	92.41	94.77	93.43	89.05	78.90
	XLNet	**94.44**	91.82	93.50	94.89	**96.29**	**91.59**	**80.35**
SFU	BERT	**86.06**	91.53	**85.40**	**85.88**	91.22	91.89	**92.39**
	RoBERTa	84.85	91.73	83.83	85.12	91.44	91.81	91.71
	XLNet	85.06	**91.81**	82.69	84.43	**92.37**	**92.42**	91.61
Negation Scope Resolution: Local								

Table 7: Results for Negation Scope Resolution (Separate Early Stopping)

Task	Dataset	Model	Author	F1
Negation Scope Resolution	BioScope Abstracts	BiLSTM-Joint	Fancellu, Lopez, Webber	92.11
		XLNet (BF+SFU) (Independently Trained)	-	96.77
		XLNet (BA+SFU) (Local) (Jointly Trained)	Ours	96.80
		XLNet (BF+BA+SFU) (Global) (Jointly Trained)	Ours	**97.06**
	BioScope Full Papers	ML MetaLearner	Morante, Daelemans	84.71
		XLNet (BA+SFU) (Independently Trained)	-	97.12
		BERT (BF+BA) (Local) (Jointly Trained)	Ours	96.78
		BERT (BF+BA) (Global) (Jointly Trained)	Ours	**97.40**
	SFU	BiLSTM	Fancellu, Lopez, Webber	89.93
		RoBERTA (BF+SFU) (Independently Trained)	-	91.34
		XLNet (BF+SFU) (Local) (Jointly Trained)	Ours	92.42
		RoBERTa (BF+BA+SFU) (Global) (Jointly Trained)	Ours	**93.19**

Table 8: Comparison of Negation Scope Resolution Results with State-of-the-Art Results

Test Dataset	Model	Train Dataset						
		BA	BA+SFU	BF	BF+BA	BF+BA+SFU	BF+SFU	SFU
BA	BERT	97.14	96.80	95.13	97.59	97.00	94.66	**82.99**
	RoBERTa	96.81	96.28	95.76	97.16	96.63	95.00	78.22
	XLNet	**97.90**	**97.68**	**96.67**	**97.90**	**97.87**	**95.69**	81.43
BF	BERT	93.16	90.87	93.78	96.22	95.07	90.49	77.67
	RoBERTa	93.50	90.90	92.31	95.39	93.20	**92.29**	75.55
	XLNet	**95.58**	**93.55**	**94.34**	96.36	95.36	91.19	**77.71**
SFU	BERT	**77.93**	89.99	**77.20**	**77.78**	91.35	90.41	89.85
	RoBERTa	75.94	**90.81**	74.78	75.76	90.86	**90.53**	89.68
	XLNet	77.82	90.31	74.54	76.47	89.98	89.89	**90.41**
Speculation Scope Resolution: Global								

Test Dataset	Model	Train Dataset						
		BA	BA+SFU	BF	BF+BA	BF+BA+SFU	BF+SFU	SFU
BA	BERT	97.09	96.48	95.20	97.36	96.48	94.68	80.68
	RoBERTa	96.67	96.44	95.01	96.94	96.68	94.68	78.64
	XLNet	**97.86**	**97.67**	**96.36**	**98.28**	**97.71**	**96.06**	**82.71**
BF	BERT	93.43	91.41	91.99	93.74	**93.52**	89.58	76.36
	RoBERTa	93.02	91.48	92.22	95.69	90.72	**91.52**	74.92
	XLNet	**94.74**	**93.87**	**94.53**	**96.04**	93.24	91.40	**78.17**
SFU	BERT	**78.10**	89.46	**77.75**	**78.33**	90.63	**90.44**	89.82
	RoBERTa	76.21	89.65	74.51	76.09	89.14	89.93	89.76
	XLNet	77.45	**90.16**	73.14	76.26	89.75	90.39	**91.17**
Speculation Scope Resolution: Local								

Table 9: Results for Speculation Scope Resolution (Combined Early Stopping)

Test Dataset	Model	Train Dataset						
		BA	BA+SFU	BF	BF+BA	BF+BA+SFU	BF+SFU	SFU
BA	BERT	97.32	96.72	95.14	97.00	96.67	94.88	81.32
	RoBERTa	96.58	97.02	95.54	96.51	96.68	93.83	76.40
	XLNet	**97.83**	**97.64**	**96.53**	**97.80**	**98.00**	**96.61**	**83.39**
BF	BERT	93.53	90.51	91.57	94.15	94.24	**92.81**	76.88
	RoBERTa	92.87	91.47	92.20	94.68	92.84	90.12	73.71
	XLNet	**95.06**	**93.32**	**94.60**	**95.83**	**95.28**	92.18	**78.64**
SFU	BERT	**78.59**	**91.08**	**77.95**	**77.88**	**90.81**	90.47	90.56
	RoBERTa	75.83	89.05	75.07	75.11	89.77	90.09	**90.64**
	XLNet	76.64	90.22	74.01	76.99	90.52	**90.85**	89.66
Speculation Scope Resolution: Global								

Test Dataset	Model	Train Dataset						
		BA	BA+SFU	BF	BF+BA	BF+BA+SFU	BF+SFU	SFU
BA	BERT	96.91	97.05	95.00	97.22	97.21	94.86	80.56
	RoBERTa	96.67	96.97	95.32	97.02	96.45	95.02	78.64
	XLNet	**98.09**	**97.72**	**96.38**	**97.43**	**97.73**	**96.14**	**80.61**
BF	BERT	93.47	92.13	93.21	94.48	92.58	90.86	76.24
	RoBERTa	93.02	91.91	90.61	94.58	94.34	91.42	74.92
	XLNet	**94.39**	**93.50**	**94.61**	**94.95**	**96.39**	**92.15**	76.49
SFU	BERT	**77.86**	89.81	**76.92**	**78.37**	89.95	89.89	**90.11**
	RoBERTa	76.21	89.60	75.55	75.28	89.42	**90.27**	89.76
	XLNet	75.97	**90.52**	73.40	73.97	**90.26**	89.57	90.10
Speculation Scope Resolution: Local								

Table 10: Results for Speculation Scope Resolution (Separate Early Stopping)

A comparison of the best models trained jointly on Negation and Speculation compared with the state-of-the-art results for Speculation Scope Resolution is shown in Table 11. Our joint training approach outperforms the existing state-of-the-art results on BioScope Abstracts and SFU Review Corpus.

5.4 Analysis

Our proposed joint training scheme clearly yields substantial improvements over the independent task-specific training approach, as we outperform the independently trained models consistently, and report new state-of-the-art results, as is illustrated in Tables 5, 8 and 11.

Task	Dataset	Model	Author	F1
Speculation Scope Resolution	BioScope Abstracts	Recursive Neural Network	Ren, Fei, Peng	93.60
		XLNet (BA) (Independently Trained)	Britto, Khandelwal	97.87
		XLNet (BF+BA) (Local) (Jointly Trained)	Ours	**98.28**
		XLNet (BF+BA+SFU) (Global) (Jointly Trained)	Ours	98.00
	BioScope Full Papers	CNN	Qian et al.	86.69
		XLNet (BF+BA) (Independently Trained)	Britto, Khandelwal	**96.91**
		XLNet (BF+BA+SFU) (Local) (Jointly Trained)	Ours	96.39
		XLNet (BF+BA) (Global) (Jointly Trained)	Ours	96.36
	SFU	SVM	Diaz, Taboada, Mitkov	78.88
		BERT (BF+SFU) (Independently Trained)	Britto, Khandelwal	91.00
		XLNet (SFU) (Local) (Jointly Trained)	Ours	91.17
		XLNet (BF+BA+SFU) (Global) (Jointly Trained)	Ours	**91.35**

Table 11: Comparison of Speculation Scope Resolution Results with State-of-the-Art Results

- XLNet consistently outperforms RoBERTa and BERT on the BioScope Corpora. For the SFU Review Corpus, we see a mixed bag of results, but BERT and XLNet tend to outperform RoBERTa. The impact of the similarity between the pretraining corpora and the dataset for which the model is finetuned could account for these observations.

Model	Negation Scope Resolution (Separate)	Speculation Scope Resolution (Separate)	Negation Scope Resolution (Combined)	Speculation Scope Resolution (Combined)
		Task		
BERT	0.57	-0.12	-0.26	-0.50
RoBERTa	0.24	0.07	0.33	-0.35
XLNet	-0.28	-0.03	-0.53	-0.08

Table 12: Difference between Local and Global Preprocessing methods for Scope Resolution

- For Scope Resolution, the Global preprocessing method tends to outperform the Local preprocessing method. This trend is visible in Table 12, which contains the difference between results using the local preprocessing method and the global preprocessing method (i.e. Local - Global), averaged across all train-test dataset combinations, shown for each model-task combination. The majority differences (8 out of 12, or 66%) are negative, showing that global preprocessing method outperforms the local preprocessing method.

- The Combined Early Stopping training method outperform the Separate Early Stopping training method. This trend is visible in Table 13, which contains the difference between the combined early stopping method

Model	Negation Cue Detection	Speculation Cue Detection	Negation Scope Resolution	Speculation Scope Resolution
	Task			
BERT	-5.48	-1.99	0.19	0.24
RoBERTa	1.89	2.36	0.02	0.38
XLNet	2.65	2.73	0.19	-0.20

Table 13: Difference between Combined and Separate Early Stopping training methodologies

and the separate early stopping method, (i.e. Combined - Separate), averaged across all train-test dataset combinations, shown for each model-task combination. The majority differences are positive, showing that Combined outperforms Separate. We reason that the combined early stopping method avoids overfitting to the validation set, due to more examples being considered in the validation set.

6 Conclusion

In this paper, we explored the realm of Multi-task training to jointly train the same model to perform both negation and speculation detection. We experimented with transformer-based architectures (BERT, XLNet and RoBERTa), and proposed schemes to jointly train the cue detection model for both negation and speculation, and the scope resolution model for both negation and speculation. Our approach yielded improvements over the independently trained versions of the same architectures, and we reported new state-of-the-art results for both negation and speculation scope resolution on the BioScope Corpus and the SFU Review Corpus. We also evaluated the different design choices that were involved, and observed that the Combined Early Stopping variant gave the best overall performance.

The future scope of this work would be to look at using this scheme to jointly train a model for more such tasks, like NER and Sentiment Analysis, along with Negation and Speculation Detection.

References

Jeremy Barnes, Erik Velldal, and Lilja Øvrelid. 2019. Improving sentiment analysis with multi-task learning of negation. *CoRR*, abs/1906.07610.

Parminder Bhatia, Busra Celikkaya, and Mohammed Khalilia. 2018. End-to-end joint entity extraction and negation detection for clinical text. *CoRR*, abs/1812.05270.

Benita Kathleen Britto and Aditya Khandelwal. 2020. Resolving the scope of speculation and negation using transformer-based architectures.

Noa P Cruz, Maite Taboada, and Ruslan Mitkov. 2016. A machine-learning approach to negation and speculation detection for sentiment analysis. *Journal of the Association for Information Science and Technology*, 67(9):2118–2136.

Jacob Devlin, Ming-Wei Chang, Kenton Lee, and Kristina Toutanova. 2018. Bert: Pre-training of deep bidirectional transformers for language understanding. *arXiv preprint arXiv:1810.04805*.

Hao Fei, Yafeng Ren, and Donghong Ji. 2020. Negation and speculation scope detection using recursive neural conditional random fields. *Neurocomputing*, 374:22–29.

Aditya Khandelwal and Suraj Sawant. 2020. Neg-BERT: A transfer learning approach for negation detection and scope resolution. In *Proceedings of The 12th Language Resources and Evaluation Conference*, pages 5739–5748, Marseille, France. European Language Resources Association.

Halil Kilicoglu and Sabine Bergler. 2008. Recognizing speculative language in biomedical research articles: A linguistically motivated perspective. *BMC bioinformatics*, 9 Suppl 11:S10.

Halil Kilicoglu and Sabine Bergler. 2010. A high-precision approach to detecting hedges and their scopes. pages 70–77.

Natalia Konstantinova, Sheila C. M. De Sousa, Noa P. Cruz, Manuel J. Maña, and Ruslan Mitkov. A review corpus annotated for negation, speculation and their scope.

Yinhan Liu, Myle Ott, Naman Goyal, Jingfei Du, Mandar Joshi, Danqi Chen, Omer Levy, Mike Lewis, Luke Zettlemoyer, and Veselin Stoyanov. 2019. Roberta: A robustly optimized bert pretraining approach. *arXiv preprint arXiv:1907.11692*.

Guillermo Moncecchi, Jean-Luc Minel, and Dina Wonsever. 2012. Improving speculative language detection using linguistic knowledge. In *Proceedings of the Workshop on Extra-Propositional Aspects of Meaning in Computational Linguistics*, pages 37–46.

Roser Morante, Vincent Asch, and Walter Daelemans. 2010. Memory-based resolution of in-sentence scopes of hedge cues. pages 40–47.

Roser Morante and Walter Daelemans. 2009. Learning the scope of hedge cues in biomedical texts. In *Proceedings of the Workshop on Current Trends in Biomedical Natural Language Processing*, BioNLP '09, page 28–36, USA. Association for Computational Linguistics.

Arzucan Özgür and Dragomir R. Radev. 2009. Detecting speculations and their scopes in scientific text. In *Proceedings of the 2009 Conference on Empirical Methods in Natural Language Processing: Volume 3 - Volume 3*, EMNLP '09, page 1398–1407, USA. Association for Computational Linguistics.

Zhong Qian, Peifeng Li, Qiaoming Zhu, Guodong Zhou, Zhunchen Luo, and Wei Luo. 2016. Speculation and negation scope detection via convolutional neural networks. In *Proceedings of the 2016 Conference on Empirical Methods in Natural Language Processing*, pages 815–825.

Jonathon Read, Erik Velldal, Stephan Oepen, and Lilja Øvrelid. 2011. Resolving speculation and negation scope in biomedical articles with a syntactic constituent ranker.

György Szarvas, Veronika Vincze, Richárd Farkas, and János Csirik. 2008. The bioscope corpus: Annotation for negation, uncertainty and their scope in biomedical texts. In *Proceedings of the Workshop on Current Trends in Biomedical Natural Language Processing*, BioNLP '08, page 38–45, USA. Association for Computational Linguistics.

Buzhou Tang, Xiaolong Wang, Xuan Wang, Bo Yuan, and Shixi Fan. 2010. A cascade method for detecting hedges and their scope in natural language text. pages 13–17.

Erik Velldal. 2011. Predicting speculation: A simple disambiguation approach to hedge detection in biomedical literature. *Journal of biomedical semantics*, 2 Suppl 5:S7.

Erik Velldal, Lilja Øvrelid, and Stephan Oepen. 2010. Resolving speculation: Maxent cue classification and dependency-based scope rules. pages 48–55.

Erik Velldal, Lilja Øvrelid, Jonathon Read, and Stephan Oepen. 2012. Speculation and negation: Rules, rankers, and the role of syntax. *Computational Linguistics*, 38:369–410.

Zhilin Yang, Zihang Dai, Yiming Yang, Jaime Carbonell, Russ R Salakhutdinov, and Quoc V Le. 2019. Xlnet: Generalized autoregressive pretraining for language understanding. In *Advances in neural information processing systems*, pages 5753–5763.

Biomedical Event Extraction
as Multi-turn Question Answering

Xing David Wang[1], Leon Weber[1,2], Ulf Leser[1]
[1]Computer Science Department, Humboldt-Universität zu Berlin
[2]Max Delbrück Center for Molecular Medicine
{wangxida, weberple, leser}@informatik.hu-berlin.de

Abstract

Biomedical event extraction from natural text is a challenging task as it searches for complex and often nested structures describing specific relationships between multiple molecular entities, such as genes, proteins, or cellular components. It usually is implemented by a complex pipeline of individual tools to solve the different relation extraction subtasks. We present an alternative approach where the detection of relationships between entities is described uniformly as questions, which are iteratively answered by a question answering (QA) system based on the domain-specific language model SciBERT. This model outperforms two strong baselines in two biomedical event extraction corpora in a Knowledge Base Population setting, and also achieves competitive performance in BioNLP challenge evaluation settings.

1 Introduction

Biomedical event extraction (BEE) (Björne and Salakoski, 2011) aims to extract molecular events from natural text, where an event typically encompasses certain biomedical entities, such as genes, proteins, complexes or cellular components, specific trigger words determining the event type, and relationships between the entities whose roles depends on the event type. For instance, the verb *phosphorylates* is a hint to a mention of a phosphorylation event in a given sentence and typically has two entities, one that controls the phosphorylation and one that is phosphorylated. Events may also involve other events, such as the inhibition of an expression, and may ultimately form partial or entire biological pathways (Gonzalez et al., 2015).

State-of-the-art methods for BEE rely on learning textual patterns and features from annotated documents where entities and their specific role in an event structure are manually marked. They typically consist of multiple classifiers to solve the different subtasks of trigger, role, and event detection, each requiring individual training and validation data. In this paper, we instead model BEE as iterative question answering, using the same model for each of the individual steps which allows knowledge sharing and joint learning of the different event components. We show that this model is as effective in predicting event structures in two BioNLP shared tasks (GENIA, 2011 and Pathway Curation, 2013) as a baseline consisting of multiple, CNN based classifiers (Björne and Salakoski, 2018).

The paper is structured as followed: In Section 2, we give a brief overview over related work in biomedical event extraction and in question answering. We define the event extraction task, our question answering model, and our evaluation setup in Section 3. In Section 4, we present our results and discuss them before we conclude the paper. The code and pretrained models are freely available at https://github.com/WangXII/bio_event_qa.

2 Related Work

Approaches to BEE can be divided into two categories: Approaches using manually defined rules (Valenzuela-Escárcega et al., 2015) and approaches making use of machine learning algorithms. Early approaches of the latter category, such as Event-Mine (Miwa and Ananiadou, 2013) or the Turku Event Extraction System (TEES) (Björne and Salakoski, 2011), had in common that they achieve event extraction through a pipeline of several independent classifiers, each solving a different subtask of event extraction and each based on a set of specifically defined features extracted from the text, often after heavy and error-prone preprocessing (e.g., POS tagging, dependency parsing). More recent

Proceedings of the 11th International Workshop on Health Text Mining and Information Analysis, pages 88–96
November 20, 2020. ©2020 Association for Computational Linguistics
https://doi.org/10.18653/v1/P17

works use neural architectures, where the previously manually defined features are replaced by automatically learned text representations (Björne and Salakoski, 2018; Trieu et al., 2020), involving techniques like word embeddings and other language models. While the original TEES (TEES SVM) (Björne and Salakoski, 2011), was based on a pipeline of SVMs using manually defined features, the more recent TEES CNN (Björne and Salakoski, 2018) additionally incorporates biomedical word embeddings as features and replaces the SVMs with CNNs. As pipelined models suffer from error propagation (for instance, an undetected event trigger in the first phase leads to missing the event entirely), approaches based on joint inference recently became more popular. Zhu and Zheng (2020) assign a separate probability to each event trigger, relation and event candidate and move the final decision about the veracity of an event structure to an optimization scheme solved in a post-processing step. DeepEventMine (Trieu et al., 2020) is a derivative of EventMine (Miwa and Ananiadou, 2013) and makes use of text representations learned by BERT (Devlin et al., 2018). It tries to avoid error propagation by training a multi-layer network for BEE in an end-to-end manner and achieves new state-of-the-art performance in various biomedical event extraction corpora. In contrast to these previous approaches, our model employs a network with only one single output layer for all event extraction subtasks and it does not need to introduce a new layer for each subtask.

In this work, we will model BEE as a an iterative question answering (QA) process. This idea was brought up first by McCann et al. (2018), who showed how to model ten different NLP tasks, among them machine translation, summarization, and sentiment analysis, as question answering tasks over a properly defined context. Li et al. (2019) proposed a specific question answering framework for event extraction based on the idea of extracting the entities of individual relations using so-called "question turns". In each turn, the question answering procedure asks a question for a new entity from the relation followed by a text passage where a span is marked as the output entity. Found entities from previous turns are included in the questions of subsequent turns to allow for more precise subsequent queries. The process is controlled by predefined question templates which determine the sequence of turns depending on the event type. However, this

work is not applicable to BEE, because it assumes a fixed number of arguments and has no support for nested events (events that serve as arguments for other events), which are two defining characteristics of the BEE task.

In this paper, we develop a similar framework for the extraction of nested biomedical events. Our framework applies SciBert (Beltagy et al., 2019), a domain-specific refinement of BERT (Devlin et al., 2018), as underlying QA method. BERT (and SciBert) is a pre-trained transformer model (Vaswani et al., 2017) which relies on an attention mechanism to learn relationships between different parts of a sequential input, which was shown to better capture long-term dependencies than Convolutional and Recurrent Neural Networks. The parameters of its final layers can be used as input features to other models, or can be used in a fine-tuning procedure involving a further, task-specific layer for problems like question answering, sentence similarity quantification, or sentence continuation prediction.

3   Material and Methods

3.1   Event Extraction

Biomedical event structures are used to model biomedical processes. In general, they consist of signal words, called trigger of the event, and biomedical entities, called arguments of the event. The trigger determines the event type, which in turn determines the semantics (or roles) of its arguments. Event triggers are often verbs or nouns such as *phosphorylation*, *transcription* or *binds*, whereas biomedical entities typically are proper nouns, such as *NF-kappa B*, *ATP* or *glucose*. The role *theme* denotes the central object of interest in an event, while the role *cause* often is the facilitator or driver of the event. Notably, events can be arguments of other events, for instance when a protein *A* (the cause) *activates* (the trigger) the phosphorylation (the theme, in this case a nested trigger) of another protein *B* (the argument of the nested event). A typical biomedical event structure is illustrated in Figure 1.

3.2   Multi-turn Question Answering

In order to find simple and complex event structures, we adopt the multi-turn question answering approach of Li et al. (2019) to BEE. We cast it as a series of QA tasks, where each individual QA problem is modeled as a sequence labeling task in

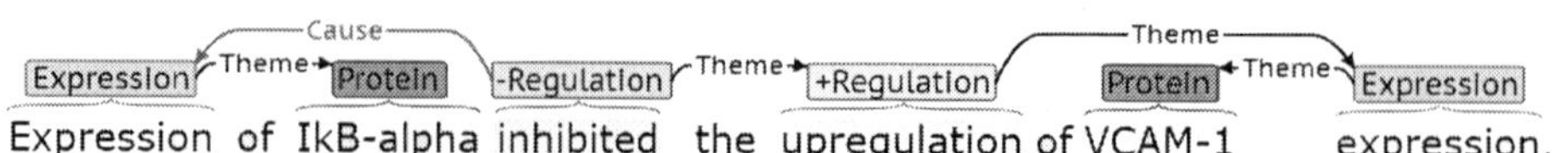

Figure 1: Event visualization using BRAT by Stenetorp et al. (2011)

Table 1: Our question template and the expected answers when applied to the example from Figure 1. In the first question we ask for simple events involving our the chosen entity as a theme. If the entity is part of an event we retrieve the corresponding event trigger, its type and position in the text. Then, we ask for other event arguments belonging to the trigger-theme pair. Subsequent questions aim to uncover recursive events containing the just extracted simple event as a theme. The recursive descent ends as soon as the event is not found to be part of another structure.

Questions:	Answers:
1. What are events of *VCAM-1*?	The Expression *expression* at (62,72).
2. What are arguments of the *Expression* of *VCAM-1*?	None.
3. What are events of the *Expression* of *VCAM-1*?	The Positive regulation *upregulation* at (39,51).
4. What are arguments of the *Positive regulation* of the *Expression VCAM-1*?	None.
5. What are events of the *Positive regulation* of the *Expression* of *VCAM-1*?	The Negative regulation *inhibited* at (25,34).
6. What are arguments of the *Negative regulation* of the *Positive regulation* of the *Expression* of *VCAM-1*?	The Cause *expression* at (1,11).
7. What are events of the *Negative regulation* of the *Positive regulation* of the *Expression* of *VCAM-1*?	None.

which the model decides for each token whether it belongs to an answer of the current question and if it does, which role it has. This can be interpreted as a kind of multitask learning in which the different tasks are not defined by different loss functions but through different types of questions. Triggers determine the specific event type whereas entities take one of the event argument roles. The formulation as sequence labeling tasks allows for multiple text spans to be tagged as answers of the same question which is beneficial as (1) an entity can participate in two distinct event structures and (2) an event can have multiple different arguments. The model assumes gold standard annotation of all entities in the corpus and uses these to structure the iterative QA process, treating each gold-standard entity as a potential theme argument. It expands events from there by iteratively asking for corresponding event triggers, event arguments and nested regulation events.

We introduce the notion of a question template which defines the different types of questions we use in our model and the sequence of turns we pose them. Our question template follows a recursive procedure and distinguishes two main question types, one for detecting event triggers and one for detecting event arguments. The process iterates through all given entities and asks whether there are any events with this entity as theme. This first question belongs to the **Triggers** question type and detects triggers corresponding to a theme candidate. In the subsequent **Arguments** question we ask for arguments belonging to a previously discovered (theme, trigger) combination. Applying the first question type **Triggers** to our example from Figure 1, we ask for all event triggers and their event type belonging to the protein *VCAM-1*. Note that this question addresses all different mentions of the entity *VCAM-1* in the given document. In our example, the assignment of answer triggers to entity evidences is clear as *VCAM-1* is mentioned exactly once in the document; in cases where an argument is mentioned more than once, we need to perform the correct assignment in a subsequent step (see next section). As the answer to our question we mark the event trigger *expression* with the event type *Expression*. In every **Arguments** question (cf. Table 1) we incorporate the event trigger found from the previous answer into the formulation of the new question. Next, we query for non-theme arguments belonging to the *Expression* of *VCAM-1* which yields no answers in this example.

The subsequent questions deal with finding nested structures and rely on the same schema of alternating **Triggers** and **Arguments** question turns. We ask which other events our previously found event could be a theme of, i.e., we ask "*Which are the events of the Expression of VCAM-1?*". In our example, we find that the *VCAM-1 expression* is *upregulated*; the trigger *upregulated* denotes an event of type *positive regulation*. If we found multiple answers of different event types to the same entity or event in a **Triggers** question, we expand each single of these into a separate event structure (see next section). In our example, we find exactly one answer to the nested trigger question and proceed again by querying for the arguments of the found event. The recursion can go on for an arbitrary amount of steps as it only stops when there is no new event trigger for a **Triggers** question[1]. In the example, we recurse twice and then stop with the result that the *upregulation* of the *VCAM-1 expression* is itself *inhibited* in a *negative regulation* event which is caused by an *expression* event.

An overview of the application of our method to the example from Figure 1 can be found in Table 1. Pseudocode of our framework is given in Algorithm 1.

We transform all the event annotations provided by the tasks to natural language questions. The mapping from event annotation to question is straightforward and not described further here.

3.3 Event Merging

The answers from our question answering model results in only basic and partly underdetermined event structures that do not fit the format of events in our evaluation corpora. We apply two different post-processing steps: Event matching, where we identify the text span best matching the prior event structure from the question and the entity/trigger from the answer, and event merging, where we merge the prior event structure and the entity/trigger from the answer into one single event structure.

We illustrate both procedures using the example from Figure 1. In the first **Triggers** questions, we receive the expression trigger at character positions (62,72) as an answer for *VCAM-1*. In the matching step, we need to identify which *VCAM-1* entity in the text the expression trigger at position (62,72) belongs to. We look up an entity and trigger dictionary, which stores positions of all entities and of all detected triggers. We then compute the differences of starting positions for each mention of the entity and the starting position of the trigger and choose the occurrence with the smallest difference. In our

example, the *VCAM-1* entity at position (55,61) is identified as a match for the trigger *expression* at position (62,72) with a difference of 7 characters. In the merging step, we combine the trigger *expression* at position (62,72) and the entity *VCAM-1* at position (55,61) to a single new event structure.

The specific algorithm for event merging depends on the question type and the possible answers. We explain the differences using two examples. Assume we found a phosphorylation event with theme A in the first **Triggers** question. Asking for arguments belonging to this prior event, assume we receive four answers, namely cause B, cause C, site D and site E. In this case of multiple argument types, we enumerate all possible cause site combinations, merge them with the prior event and receive four new phosphorylation events, i.e., phosphorylation of theme A with cause B and site D, phosphorylation of theme A with cause B and site E etc. Details regarding the performance for this merging heuristic is found in Table 4, query five. A more sophisticated merging approach is needed for binding and pathway events which may contain multiple participants. For these events, we store a directed graph per event trigger where nodes are participants and a directed edge exists from entity A to entity B if B is answer to the **Arguments** question of A. After the graph is constructed, we transform it into an undirected graph where we keep all edges which exist in both directions. In the final step, we detect maximal cliques in the graph and form a distinct binding/pathway event for each clique. The results for binding/pathway merging is found in Table 4, query six. We use similar heuristics for the merging step of nested regulations and other event types.

3.4 Implementation

We use Huggingface's Transformers[2] (Wolf et al., 2019) library in Pytorch for our implementation. For the initialization of the pretrained BERT neural network model we use SciBERT[3] (Beltagy et al., 2019) which has been pretrained on scientific literature. We add one softmax layer as output on top of the final hidden representation of each token as we fine-tune the model parameters for our question answering task. In the final output layer each token in a given document sequence is tagged in IOB2-style as either being inside, outside or the beginning of

---

[1]The deepest nesting occurring in our two evaluation corpora is three.

[2]https://github.com/huggingface/transformers
[3]https://github.com/allenai/scibert

Algorithm 1 Pseudocode of our QA framework for the extraction of event structures. We expand event structure candidates around potential theme arguments, adding corresponding event triggers in the question **Triggers** and corresponding event arguments in the question **Arguments**. If we have found new events we add their (theme, trigger)-pair to our event candidates list for the next iteration, where we ask whether the just found event is a theme to a (new) nested event.

```
1: event_candidates = proteinsFromDocument()
2: while event_candidates ≠ ∅ do
3:     new_events = ∅
4:     for candidate in event_candidates do
5:         new_triggers = Triggers(candidate)
6:         new_arguments = Arguments(candidate)
7:         new_events.add(Event(candidate, new_triggers))
8:     end for
9:     event_candidates = new_events
10: end while
```

an answer token. The beginning and inside tags are further divided into the different event type and event argument classes according to the structures seeked in a corpus. The same BERT neural network model is shared across the whole task and all questions. This allows knowledge sharing and joint learning of the different questions.

For training, we create all existing questions in the training set exactly once in the beginning and then draw randomized batches as our training examples. We use the default AdamW configuration with learning rate 5e-5, no weight decay, $\beta_1 = 0.9$, $\beta_2 = 0.999$ and $\epsilon = 1e\text{-}8$. Training is conducted on four Nvidia GeForce RTX 2080 Ti GPUs. Our maximum sequence length for the input data is 384 tokens. To deal with longer sequences than the maximum sequence length, we duplicate the beginning and the end of intermediate sequences so that they form overlapping windows with a length of 64 tokens. To decide between two differently predicted tags for the same token in two adjacent windows, we choose the tag of the token which has the larger context window. We enable apex[4] fp16 16-bit mixed precision for improved computation efficiency.

Hyperparameters to choose are the batch size and the number of epochs when to stop training. During our model development, a batch size of 16 has proven to work well together with 16 epochs after which the validation loss usually does not improve anymore. The whole training process during fine-tuning is relatively fast and the training time ranges from half an hour to an hour on Pathway Curation to around two hours on GENIA depending on hyperparameter choice. Performance in evaluation fluctuates over few percentage in F1-score

---

[4]https://github.com/NVIDIA/apex

depending on the initial seed during neural network initialization. As we mainly compare to Björne and Salakoski (2018), we adopt their evaluation strategy and report the results of the seed with the best performance on the validation set.

3.5 Corpora

We evaluate our approach to BEE on two corpora used widely in biomedical NLP research, namely the Pathway Curation corpus (PC) from the BioNLP13 challenge (Ohta et al., 2013) and the Genia 11 corpus (GENIA) from the BioNLP11 challenge (Kim et al., 2011). These corpora consist of annotated PubMed abstracts and full texts. The PC dataset focuses on pathway reactions whereas GENIA aims to cover molecular biology in general. GENIA contains 14,958 sentences and PC 5,040 sentences. GENIA distinguishes seven different event types and six different argument types, whereas PC distinguishes 24 different event types and nine argument types. Both corpora include common biochemical event types, such as *phosphorylation*, *gene expression*, *binding* or *positive (negative) regulation*. PC further distinguishes multiple conversion types, such as *dephosphorylation*, *acetylation*, *ubiquitination* etc. and it adds *activation* and *inactivation* to the class of regulation events. PC also annotates event modifiers, i.e., *speculation* and *negation*, and allows for events without a theme. The latter two types of event components currently are not addressed by our work, but could be included by adding further turns and questions to our question template. A closer breakdown of the events and their components in the two corpora can be found in Table 2.

Table 2: Statistics of our question answering training datasets built from the gold event annotations.

Question type	GENIA11		Pathway Curation	
	#questions	#gold answers	#questions	#gold answers
Simple Events				
Triggers	6,392	6,549	4,316	3,857
Arguments	6,263	1,486	3,242	2,389
Nested Events				
Triggers	10,564	3,523	5,012	1,708
Arguments	4,303	1,096	1,775	1,440
Total	27,522	12,654	14,345	9,394

3.6 Evaluation Tasks

We evaluate our model for two different tasks: Knowledge Base Population (KBP) and the standard BioNLP a* setting. In both cases, gold-standard entity annotations are provided with the corpus whereas event annotations have to be predicted.

Knowledge Base Population

Following Kim et al. (2015), we evaluate the models' capability to answer a set of predefined queries, such as finding all pairs of proteins that bind to each other. An overview of the different knowledge base queries is found in Table 3. The first four queries can be directly answered from our question answering model while the remaining three require event merging, which we perform as described in Section 3.3. As usual in KBP settings, the extracted event structures are compared on a document-level, so a same event occurring twice in a single document is counted once only in this format.

BioNLP .a* evaluation

The .a* evaluation format is the standard evaluation format provided by the GENIA and PC shared tasks. PC is conducted in a strict matching evaluation mode, where the extracted triggers, all event arguments, and their text spans must exactly coincide. The approximate span and approximate recursive matching mode for GENIA is more lenient as the text spans and positions may differ up to one word from the gold-standard annotations and nested regulation events only need to coincide in their theme arguments.

4 Results and Discussion

4.1 Knowledge Base Population

We use TEES SVM (Björne and Salakoski, 2011) and TEES CNN (Björne and Salakoski, 2018) [5] as baselines for knowledge base population. Both provide result files and models online[6]. We compare the result of our single homogeneous QA multi-turn model to the individual models of these approaches.

The results can be found in Table 4. Our approach achieves a 0.87 percentage points (pp) and a 2.47 pp better F1-score than TEES CNN and TEES SVM, respectively, on GENIA. On PC, it achieves a 2.40 pp and a 3.13 pp better F1-score. This increase can be attributed to a considerably better recall (2.35 pp for GENIA and 6.59 pp for PC, compared to TEES CNN). Its precision is 1.38 pp and 2.24 pp lower than the respective best baseline result. It shows performance gains of up to 5.16 pp F1 in the first three *Basic Event* queries which require no event merging. Results for the other type of queries are mixed: Our model achieves good results for binding and pathway pairs, yet is worse for transitive protein regulations and the combination of all conversion arguments.

Most likely, the question answering approach achieves strong performances in extraction of simple events as they rely on only one or two questions and require no complicated merging steps. The model infers binding and pathway pairs in the fifth query relatively well since we explicitly query for those in the **Arguments** question type. The worse results for the arguments of a conversion event in the sixth query are probably due to the naive heuristic of simply enumerating all valid argument combinations as output during event merging. Regulation event detection in the forth and seventh query presumably also suffer from our too-simple event merging as we match a detected event trigger cause to a whole previously discovered event structure. We also observe that error propagation negatively influences regulation detection and event detection as we immediately extract simple events after our first **Triggers** question from the (theme, trigger)-pairs, but we do not incorporate event arguments or regulations found in later question turns into a joint extraction of events.

[5] https://github.com/jbjorne/TEES
[6] https://b2share.eudat.eu/records/
bee50aa63b0b404da9c76b29de4d8653

Table 3: Queries for our Knowledge Base Population evaluation, adapted from Kim et al. (2015). We conduct evaluation of found events at document level, i.e., counting unique event structures per document. The answers are denoted as tuples. Example questions and answers are given in italics.

Knowledge Base Queries on a document	
Query description	Example answer
1. Which protein appears in context of event *A*?	(EventType, ProteinTheme)
- Which protein appears in context of a gene expression?	*- (Gene Expression, MACS1)*
2. What is an argument of event *A* of entity *X*?	(EventType, ProteinTheme, ArgumentType, Argument)
- What is the location of the localization of MACS1?	*- (Localization, MACS1, ToLoc, mitochondrial matrix)*
3. Is the simple event *A* part of a regulation?	(SimpleEvent, Boolean)
- Is the transport of hydroxyl part of a regulation?	*- ((Transport, hydroxyl), yes)*
4. What regulates the simple event *A*?	(SimpleEvent, Cause)
- What regulates the transport of hydroxyl?	*- ((Transport, hydroxyl), amiloride)*
5. What is the site of the conversion event of *A* with cause *B*?	(EventType, ProteinTheme, ProteinCause, ProteinSite)
- What is the site for the acetylation of H3 by Asf1?	*- (Acetylation, H3, Asf1, K56)*
6. What binds to protein *A*?	(Protein1, Protein2)
- What binds to Na+?	*- (Na+, H+)*
7. What regulates *A* transitively?	(ProteinTheme, ProteinCause)
- What regulates NF-kappaB?	*- (NF-kappaB, TLR2)*

Table 4: Results for Knowledge Base Population on the development sets, compared to TEES SVM and TEES CNN. Semantics for each individual question are found in Table 3. The answers of the first four queries (Simple Events) can be derived by our model without event merging. The two lower sections show only F1 scores. The best value in each partial column is marked in bold.

	GENIA				Pathway Curation			
Metric/Question type	TEES SVM	TEES CNN	QA with BERT	Support	TEES SVM	TEES CNN	QA with BERT	Support
F1 (Total)	59.78	61.38	**62.25**	3625	56.06	56.79	**59.19**	3141
Precision (Total)	68.80	**69.68**	68.30	3625	**60.57**	60.52	58.33	3141
Recall (Total)	52.86	54.84	**57.19**	3625	52.18	53.49	**60.08**	3141
1. Theme Trigger Pairs	73.07	75.23	**79.41**	1301	69.21	69.34	**74.50**	866
2. Event Arguments	**49.17**	46.76	47.36	568	45.84	46.94	**49.31**	648
3. Nested Regulation Events	63.61	66.40	**71.08**	585	66.14	64.43	**71.05**	339
4. Nested Regulation Causes	39.71	**44.21**	36.03	384	**46.19**	43.78	44.44	419
Basic Events (Total)	63.24	64.38	**66.53**	2838	58.21	57.84	**61.27**	2272
5. Full Conversion Events	-	-	-	0	38.89	**61.11**	56.25	16
6. Binding/Pathway Pairs	55.14	46.03	**60.18**	126	56.84	53.64	**60.59**	138
7. Transitive Regulations	42.14	**50.05**	38.64	660	48.72	**53.79**	51.14	715
Merged Events (Total)	44.58	**49.38**	42.80	787	50.00	**53.93**	53.20	869

Table 5: Results on the standard .a* evaluation of BioNLP shared tasks, comparing our model with four competitors. The test set evaluation is conducted online where predictions are submitted to a server and the final results are returned. DeepEventMine (Trieu et al., 2020) represents results of very recent work. Note that our model does not account for event modifications or events without themes in the PC corpus. Dev (adjusted) denotes the results on the PC development set excluding these annotations. The best value in each partial column is marked in bold.

	GENIA11				Pathway Curation				
		Test Set		*Dev Set*		Test Set		*Dev Set*	*Dev (adjusted)*
Task/Data set	F1	Precision	Recall	*F1*	F1	Precision	Recall	*F1*	*F1*
TEES SVM (Björne and Salakoski, 2011)	53.30	57.65	49.56	*56.00*	51.10	55.78	47.15	*44.34*	*45.55*
EventMine (Miwa and Ananiadou, 2013)	57.98	63.48	53.35	-	52.84	53.48	**52.23**	-	-
TEES CNN (Björne and Salakoski, 2018)	56.80	64.86	50.53	*58.57*	52.10	58.31	47.08	*46.07*	*47.10*
DeepEventMine (Trieu et al., 2020)	**63.02**	**71.71**	56.20	*62.75*	**55.67**	**64.12**	49.19	*56.57*	-
QA with BERT	58.33	59.33	**57.37**	*56.50*	48.29	48.74	47.85	*44.60*	***47.59***

4.2 BioNLP .a* Evaluation

In Table 5, evaluation results in the BioNLP .a* challenge setting are compared to four competitors: TES CNN (Björne and Salakoski, 2018), TEES SVM (Björne and Salakoski, 2011), EventMine (Miwa and Ananiadou, 2013), and the very recent DeepEventMine (Trieu et al., 2020). On GENIA11, our proposed approach beats three competitors on the test set, but is outperformed by DeepEvent-Mine by almost 5 pp in F1-score. The higher recall and lower precision compared to DeepEventMine might be attributed to the simple rule-based event merging step, which constructs events for all detected relations regardless of their score. In contrast, DeepEventMine models the event construction as a separate machine learning task in which errors from earlier steps can be corrected, potentially leading to a higher precision.

For the PC corpus, our results are considerably worse than those of the baselines on both the dev and the test set. This inferior performance can be attributed to the fact that the proposed model does not account for event modifications or events without themes. Accordingly, we evaluated the models again on the development set excluding such annotations. The results for this experiment can be found in the column *Pathway Curation Dev (adjusted)*. Under this setting our proposed model outperforms both TEES variants. Note that events without themes and their regulations make up to a tenth of the events in the development set of PC, among them the majority are simple pathway events only made up by an event trigger.

4.3 Error Analysis

Table 6: Error statistics of our question answering model.

Error type	GENIA11		Pathway Curation	
	# wrong answer	%	#questions	%
Wrong Trigger Spans	643	46.5	912	67.8
Wrong Trigger Label	56	4.1	60	4.4
Wrong Argument Spans	674	48.9	370	27.5
Wrong Argument Label	7	0.5	3	0.2
False Positives (Total)	1,380	100	1,345	100
Missing Trigger Spans (Question)	335	31.8	642	40.1
Missing Trigger Spans (Propagated)	122	11.6	243	15.2
Wrong Trigger Label	56	5.3	60	3.7
Missing Argument Spans (Question)	177	16.9	194	12.1
Missing Argument Spans (Propagated)	356	33.4	458	28.7
Wrong Argument Label	7	0.7	3	0.2
False Negatives (Total)	1,053	100	1,600	100

We conducted an error analysis on the dev sets of the GENIA11 and PC corpora. Results are shown in Table 6. We distinguish error types into false positives and false negatives:

- *Wrong Trigger/Argument Spans* denotes answers predicted by the model which are no gold-standard answers.

- *Wrong Trigger/Argument Label* means correctly detected text spans which have the wrong event type or wrong argument type.

- *Missing Trigger/Argument Spans (Question)* refers to questions where a trigger or an argument has not been extracted.

- *Missing Trigger/Argument Spans (Propagated)* refers to triggers or arguments which have not been extracted because the according question has not been found (i.e., the answers from a previous question have been wrong so that the subsequent question is not posed).

We find that wrong label assignment is the cause for about five percent of false positives and false negatives. Missing propagated questions make up about one half of the false negatives during question answering in non-regulation event types. The relative amount of errors is lower in GENIA11 compared to Pathway Curation which reflects the overall better model performances in GENIA11.

In an ablation study, we examine the impact of joint training on all questions versus training only on the one question type of simple events trigger detection, i.e., only using the examples of the first **Triggers** question and examining the impact of multi-task learning in our model. We find that training the model only on the one question type results in a worse performance (1.08 pp F1-score) for answering this one specific question compared to evaluating the found triggers trained on the full questions dataset. This indicates that the shared model parameters provide a benefit for detecting the right answer to all question types.

5 Conclusion

We presented an approach for BEE in which this task is modeled as multi-turn question answering problem using BERT as underlying language model. We show that our model is able to form event structures from the answers of multiple questions. Our experiments show promising results on two corpora, especially in a Knowledge Base Population setting. In future work, we aim to improve model performance by adjusting the event merging procedure and by using further or modified question templates. It would also be worthwhile to study the reasons of the performance gains of our model compared to TEES in more detail, for instance by replacing the CNN in TEES CNN with BERT.

References

Iz Beltagy, Kyle Lo, and Arman Cohan. 2019. Scibert: Pretrained language model for scientific text. In *EMNLP*.

Jari Björne and Tapio Salakoski. 2011. Generalizing biomedical event extraction. In *Proceedings of BioNLP Shared Task 2011 Workshop*, pages 183–191.

Jari Björne and Tapio Salakoski. 2018. Biomedical event extraction using convolutional neural networks and dependency parsing. In *Proceedings of the BioNLP 2018 workshop*, pages 98–108.

Jacob Devlin, Ming-Wei Chang, Kenton Lee, and Kristina Toutanova. 2018. Bert: Pre-training of deep bidirectional transformers for language understanding. *arXiv preprint arXiv:1810.04805*.

Graciela H Gonzalez, Tasnia Tahsin, Britton C Goodale, Anna C Greene, and Casey S Greene. 2015. Recent advances and emerging applications in text and data mining for biomedical discovery. *Briefings in bioinformatics*, 17(1):33–42.

Jin-Dong Kim, Jung-jae Kim, Xu Han, and Dietrich Rebholz-Schuhmann. 2015. Extending the evaluation of genia event task toward knowledge base construction and comparison to gene regulation ontology task. *BMC bioinformatics*, 16(S10):S3.

Jin-Dong Kim, Yue Wang, Toshihisa Takagi, and Akinori Yonezawa. 2011. Overview of genia event task in bionlp shared task 2011. In *Proceedings of the BioNLP Shared Task 2011 Workshop*, pages 7–15. Association for Computational Linguistics.

Xiaoya Li, Fan Yin, Zijun Sun, Xiayu Li, Arianna Yuan, Duo Chai, Mingxin Zhou, and Jiwei Li. 2019. Entity-relation extraction as multi-turn question answering. In *Proceedings of the 57th Annual Meeting of the Association for Computational Linguistics*, pages 1340–1350.

Bryan McCann, Nitish Shirish Keskar, Caiming Xiong, and Richard Socher. 2018. The natural language decathlon: Multitask learning as question answering. *arXiv preprint arXiv:1806.08730*.

Makoto Miwa and Sophia Ananiadou. 2013. Nactem eventmine for bionlp 2013 cg and pc tasks. In *Proceedings of the BioNLP Shared Task 2013 Workshop*, pages 94–98.

Tomoko Ohta, Sampo Pyysalo, Rafal Rak, Andrew Rowley, Hong-Woo Chun, Sung-Jae Jung, Sung-Pil Choi, Sophia Ananiadou, and Jun'ichi Tsujii. 2013. Overview of the pathway curation (pc) task of bionlp shared task 2013. In *Proceedings of the BioNLP Shared Task 2013 Workshop*, pages 67–75.

Pontus Stenetorp, Goran Topić, Sampo Pyysalo, Tomoko Ohta, Jin-Dong Kim, and Jun'ichi Tsujii. 2011. Bionlp shared task 2011: Supporting resources. In *Proceedings of BioNLP Shared Task 2011 Workshop*, pages 112–120, Portland, Oregon, USA. Association for Computational Linguistics.

Hai-Long Trieu, Thy Thy Tran, Khoa NA Duong, Anh Nguyen, Makoto Miwa, and Sophia Ananiadou. 2020. Deepeventmine: End-to-end neural nested event extraction from biomedical texts. *Bioinformatics*.

Marco A Valenzuela-Escárcega, Gus Hahn-Powell, Mihai Surdeanu, and Thomas Hicks. 2015. A domain-independent rule-based framework for event extraction. In *Proceedings of ACL-IJCNLP 2015 System Demonstrations*, pages 127–132.

Ashish Vaswani, Noam Shazeer, Niki Parmar, Jakob Uszkoreit, Llion Jones, Aidan N Gomez, Łukasz Kaiser, and Illia Polosukhin. 2017. Attention is all you need. In *Advances in neural information processing systems*, pages 5998–6008.

Thomas Wolf, Lysandre Debut, Victor Sanh, Julien Chaumond, Clement Delangue, Anthony Moi, Pierric Cistac, Tim Rault, Rémi Louf, Morgan Funtowicz, et al. 2019. Transformers: State-of-the-art natural language processing. *arXiv preprint arXiv:1910.03771*.

Lvxing Zhu and Haoran Zheng. 2020. Biomedical event extraction with a novel combination strategy based on hybrid deep neural networks. *BMC bioinformatics*, 21(1):47.

An efficient representation of chronological events in medical texts

Andrey Kormilitzin[1,*] **Nemanja Vaci**[2,*] **Qiang Liu**[1], **Hao Ni**[3,5],
Goran Nenadic[4,5] and **Alejo Nevado-Holgado**[1,6]

[1]Department of Psychiatry, University of Oxford, Oxford, OX3 7JX, UK
[2]Department of Psychology, University of Sheffield, Sheffield, S1 1HD, UK
[3]Department of Mathematics, University College London, London, WC1H 0AY, UK
[4]School of Computer Science, University of Manchester, Manchester, M13 9PL, UK
[5]The Alan Turing Institute London, London, NW1 2DB, UK
[6]Akrivia Health, Oxford, OX1 1BY, UK
`andrey.kormilitzin@psych.ox.ac.uk`

Abstract

In this work we addressed the problem of capturing sequential information contained in longitudinal electronic health records (EHRs). Clinical notes, which is a particular type of EHR data, are a rich source of information and practitioners often develop clever solutions how to maximise the sequential information contained in free-texts. We proposed a systematic methodology for learning from chronological events available in clinical notes. The proposed methodological *path signature* framework creates a non-parametric hierarchical representation of sequential events of any type and can be used as features for downstream statistical learning tasks. The methodology was developed and externally validated using the largest in the UK secondary care mental health EHR data on a specific task of predicting survival risk of patients diagnosed with Alzheimer's disease. The signature-based model was compared to a common survival random forest model. Our results showed a 15.4% increase of risk prediction AUC at the time point of 20 months after the first admission to a specialist memory clinic and the signature method outperformed the baseline mixed-effects model by 13.2 %.

1 Introduction

Electronic health records (EHRs) have now become ubiquitous and offer novel opportunities for clinical research by supporting the development of intelligent decision support systems and improvement of patients' care. One of the distinct features of EHR is that the data are being collected over time and might be seen as health data streams, allowing research to study longitudinal trends and make inference about the progression of disease, treatments and outcomes. However, the proper representation of sequential medical events still remains

a challenge. Moreover, longitudinal clinical notes exhibit a multi-level hierarchical structure, where events are described and embedded in sentences, sentences in paragraphs and eventually resulting in chronologically ordered documents. Recent works have addressed the problem of capturing this information directly from raw texts by introducing novel neural network architectures, such as attention-based recurrent neural networks (Bai et al., 2018) and time-aware Transformers (Zhang et al., 2020). When dealing with chronological clinical notes, practitioners make multiple decisions on how to structure and transform these sequential events, which are often simplifications of medical histories. In this work we proposed a different methodology to address the problem of learning from events found in clinical notes, by first extracting them using natural language processing and then representing the sequential order by means of the *path signatures*. The signature (Lyons, 2014) is a non-parametric representation of heterogeneous sequential data, offers a feature extraction method from longitudinal events and can naturally be integrated within a general data mining pipeline. To demonstrate the methodology, we used the largest secondary care mental health EHR data in the UK to develop a survival prognostic model for patients diagnosed with Alzheimer's disease.

2 Method

2.1 Data

The data in this study were sourced from the UK-Clinical Record Interactive Search system (UK-CRIS), which provides a research platform (https://crisnetwork.co/) for data mining and analysis using de-identified real-world observational electronic patients records from twelve secondary care UK Mental Health NHS Trusts (Goodday et al., 2020). UK-CRIS provides access to struc-

---

[1]Equal contribution.

Proceedings of the 11th International Workshop on Health Text Mining and Information Analysis, pages 97–103
November 20, 2020. ©2020 Association for Computational Linguistics
https://doi.org/10.18653/v1/P17

tured information, such as ICD-10 coded diagnoses, quality of life scales and demographic information, as well as various unstructured texts, such as clinical summaries, discharge letters and progress notes. The study cohort jointly comprised records from 24,108 patients diagnosed with Alzheimer's disease and various types of dementia, containing more than 3.7 million individual clinical documents from two centres: Oxford and Southern Health Foundation NHS Trusts. The field of clinical NLP in general, and of mental health and Alzheimer's research in particular, largely suffers from the dearth of gold-annotated data. The reason is due to the shortage of trained annotators with clinical background who are also authorised to access sensitive patient-level data. Therefore, to develop a robust information extraction (IE) model from an insufficient amount of data, we leveraged the idea of transfer learning using the publicly available MIMIC-III corpus (Johnson et al., 2016) comprising information relating to patients admitted to intensive care units (ICU) with more than 2.1 million clinical notes as well as 505 gold-annotated by clinical experts discharge summaries from the 2018 n2c2 challenge (Henry et al., 2020). We assert that the study was independently approved and granted by the Oxfordshire and Southern Health NHS Foundation Trust Research Ethics Committees.

2.2 Information extraction model

The information extraction model was developed to identify diagnosis, medications and cognitive health assessment Mini-Mental State Examination score (MMSE) (Pangman et al., 2000). Additionally, the identified entities were classified according to several attributes, such as the 'experiencer' modality (i.e., whether the MMSE was actually referring to a patient or to a family member), temporal information (i.e the date of diagnosis or MMSE score) and negations (i.e. discontinued medications) (Harkema et al., 2009). Such drug mentions were discarded in order to extract the most accurate information. Generic and brand drug names were normalised using the British National Formulary, the core pharmaceutical reference book (Committee et al., 2019). The architecture of the named entity recognition model comprised a hybrid approach of an ontology-based fuzzy pattern matching and a bi-directional LSTM neural network architecture with the attention mechanism (Bahdanau et al., 2014) for sequence classification.

The GloVE word embedding (Pennington et al., 2014) were fine-tuned on both MIMIC-III and UK-CRIS data (Vaci et al., 2020; Kormilitzin et al., 2020). The developed IE model was trained only on data from the Oxford Health NHS Trust instance and externally validated on a sample of data from a regionally different Southern Health NHS Foundation Trust.

2.3 The signature of a path

Repeated measurements, speech, text, time-series or any other sequential data might be seen as a path-valued random variable. Formally, a path X of finite length in d dimensions can be described by the mapping $X : [a, b] \rightarrow \mathbb{R}^d$, or in terms of co-ordinates $X = (X_t^1, X_t^2, ..., X_t^d)$, where each coordinate X_t^i is real-valued and parametrised by $t \in [a, b]$. The signature representation S of a path X is defined as an infinite series:

$$S(X)_{a,b} = (1, S(X)_{a,b}^1, S(X)_{a,b}^2, ..., S(X)_{a,b}^d,$$
$$S(X)_{a,b}^{1,1}, S(X)_{a,b}^{1,2}, ...),$$

$$(1)$$

where each term is a k-fold iterated integral of the path X labelled by multi-index $i_1, ..., i_k$:

$$S(X)_{a,b}^{i_1,...,i_k} = \int_{a<t_k<b} ... \int_{a<t_1<t_2} \mathrm{d}X_{t_1}^{i_1}...\mathrm{d}X_{t_k}^{i_k}.$$

$$(2)$$

However, in many real-life applications the first k-terms of the truncated signature at level L give a satisfying approximation. Intuitively, it is analogous to statistical moments of a d-dimensional vector-valued random variable, such as mean, variance or higher moments. One can define statistical moments of a *path*-valued random variable, which are essentially the *signature* moments (Chevyrev and Oberhauser, 2018) defined in Eq. (2). The signature $S(X)$ completely characterises a path X up to tree-like equivalence and is invariant to reparameterisation (Hambly and Lyons, 2010). The signature can also be expressed in a more compact form known as *log-signature* (Liao et al., 2019; Morrill et al., 2020a), which is the formal power series of $\log S(X)$, while carrying the same information. Informally, the path signature captures the order of events. For example, consider two sequences $X_1 = aabba$ and $X_2 = baaab$ consisting of a simple vocabulary with only two letters $\{a, b\}$. The sequences might be presented as paths in 2d space as shown in Fig. 1. Each linear segment between

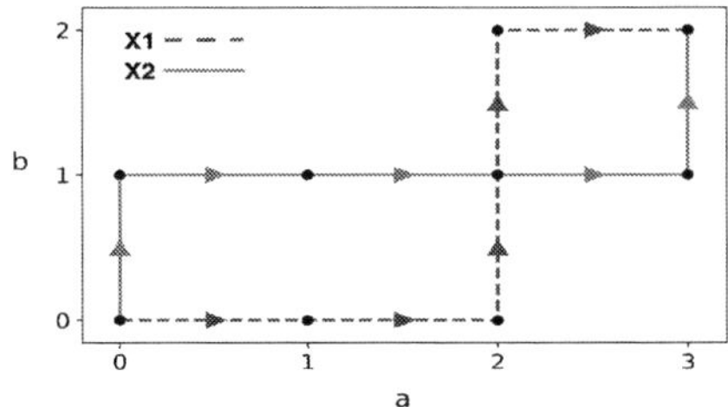

Figure 1: Two paths $X_1 = aabba$ and $X_2 = baaab$.

two points (Fig. 1) corresponds to a single letter in the sequence and the arrows denote the temporal direction of the sequence. Despite the same

Level	1		2	3		4		
$S(X_1)$	3	2	1	-0.5	-1	-1/3	-0.5	0
$S(X_2)$	3	2	0	1.5	0.5	0	0	0

Table 1: The first $k = 8$ terms of the log signature expansion up to level $L = 4$. The difference between two sequences X_1 and X_2 is apparent starting from the second level.

number of letters in the sequences $\{a = 3, b = 2\}$, the order of letters matters. The signature easily picks the differences and the first four levels of the log-signatures of paths are shown in Table 1. The lower order signature terms $S^{(i)}$ are the increments along the i-th direction (i.e. the distance between the endpoints), for example, $S^{(1)} = 3 - 0 = 3$ and $S^{(2)} = 2 - 0 = 2$ as can be seen in Figure 1. The second order corresponds to the area enclosed by a path and a chord connecting endpoints (Chevyrev and Kormilitzin, 2016).

The usefulness of a path signature as a feature map of sequential data was demonstrated theoretically (Chevyrev and Oberhauser, 2018) as well as in numerous machine learning applications in healthcare (Morrill et al., 2019; Kormilitzin et al., 2016; Arribas et al., 2018; Morrill et al., 2020b; Kormilitzin et al., 2017), finance (Arribas, 2018), computer vision (Yang et al., 2017; Xie et al., 2017), topological data analysis (Chevyrev et al., 2018) and deep learning (Kidger et al., 2019).

2.4 Independent and outcome variables

The independent variables used in the prognostic model were medications and the MMSE scores collected over time. The dependent outcome variable was right-censored time to death data in months. A synthetic example of the patient's records (Table 2) and the corresponding algorithmically extracted longitudinal data is presented in Table 3.

The outcome variable was encoded as a tuple: $(True, 34.17)$ indicating that a person has died after 34.17 months since the very first visit to a specialist memory clinic. The patient was treated by two different medications with a changing pattern and eventually was tapered off medication due to no further expected improvement.

2.5 Baseline longitudinal data summarisation

The signature transformation might be seen as a hierarchical statistical summarisation ("feature extraction") of the longitudinal data along the temporal dimension. In order to benchmark the proposed method, we used a time-honoured linear mixed-effects regression as a baseline model for longitudinal summarisation. Specifically, each patient-level longitudinal MMSE scores were modelled using a linear regression and the resulting coefficients, such as an intercept and a slope, were used as features representing the progression of the MMSE over time. The median number of medication categories was used as an additional feature, resulting in three features for each patient.

2.6 Survival random forests

The common statistical approach to analyse the time-to-event survival data is based on the linear Cox model (Collett, 2015). However, Miao et al. (2015) showed that a survival random forest (SRF) approach (Ishwaran et al., 2008) outperformed linear Cox model, based on the Harrell's concordance index (C-index) (Harrell et al., 1982), and was understandably capable of identifying non-linear effects of the input variables as opposed to linear Cox model. Therefore, we chose the SRF as the preferred method. The SRF approach was implemented in Python using "scikit-survival" package (Pölsterl et al., 2015). The Harell's C-index (the concordance index) is a goodness of fit measure for risk scores models. It is a common statistical approach to evaluate risk models in survival analysis, where data may be right-censored and corresponds to rank correlation between predicted risk scores and observed time points, similarly to Kendall's τ.

3 Results

3.1 Information extraction model

We used a hybrid approach to developing an IE model consisting of training a baseline model using MIMIC-III and n2c2 annotated data. Specifically, the named-entity recognition (NER) model

Doc date	Text
05-Oct-2016	*Today I saw a patient diagnosed with Alzheimer's, who deteriorated: MMSE 23/30 as compared to 25/30 from 1st January. Started on Rivastigmine.*
12-Feb-2017	*Today MMSE 19, the patient didn't respond to Rivastigmine and was changed to Donepezil.*
03-Feb-2018	*Great response to new treatment (MMSE 23/30), continue on Donepezil.*
01-Apr-2019	*The patient stopped responding to Donepezil and severely deteriorated (MMSE 14/30), stop Donepezil.*

Table 2: A synthetic example of chronological medical records.

Date	Medication	MMSE
01-Jan-2016	NoMed	25/30
05-Oct-2016	Rivastigmine	22/30
12-Feb-2017	Donepezil	19/30
03-Feb-2018	Donepezil	23/30
01-Apr-2019	Discontinued	14/30

Table 3: Extracted and chronologically structured data from Table 2.

comprised a transition-based system based on the chunking model (Lample et al., 2016) where tokens were represented as hashed and embedded representations of the prefix, suffix, shape and lemmatised features of words, followed by the rule-based matching using the BNF vocabulary. The IE model was implemented using "spaCy" python library[1], including negations and temporal information identification as well as relationships classification between the word-tokens using linguistic features, such as part-of-speech and dependencies. Finally, the active learning tool "Prodigy"[2] was used for iterative model improvement. Target domain training, validation and external validation data contained a collection of gold-annotated drug names, diagnosis and cognitive health assessment MMSE scores as shown in Table 4. The IE model

Concept	Training	Validation	External val.	Total
Drug	216	153	30	399
MMSE	169	87	23	279
Diagnosis	570	352	26	948

Table 4: The number of gold-annotated instances in the training, validation and external validation data sets.

achieved a good and consistent performance on both validation and external validation data sets (Table 5). The annotation schema was developed following the recommendations of Pustejovsky and Stubbs (2012). The token-level performance metrics were evaluated using the SemEval schema (Segura Bedmar et al., 2013) and the inter-annotator agreement (IAA) of two clinical annotators was

[1] https://spacy.io
[2] https://prodi.gy

computed using F1 score.

Concept	Validation			External val.			IAA
	Pr	Re	F1	Pr	Re	F1	F1
Diagnosis	89.6	96.3	92.8	84.1	86.3	84.8	95
Drug	98	98	98.1	92.4	68.4	78.3	96
MMSE	92.6	74.7	82.8	85.6	81.2	82.6	100

Table 5: Performance (shown in %) of the information extraction model. IAA - inter annotator agreement.

	n	male	female	survival time
died	1962	841	1121	52.2(22.8)
censored	1500	529	971	28.4(16.6)

Table 6: Summary statistics of the extracted data for survival analysis. Survival time is shown as mean(std) in months. The MMSE scores were not observed for censored people later in time, while date of death was recorded in hospital.

3.2 Prognostic model

Four prognostic models were developed and compared to each other. All models estimated the survival probability of a patient diagnosed with Alzheimer's disease since their first admission to a memory clinic. We compared signature ("Sig", Sec. 2.3) versus non-signature ("Non-sig", Sec. 2.5) models. We also estimated the added value of the sequential information contained in the treatment course with medications. Specifically, we used two sets of input variables: {time, MMSE} and {time, MMSE, medications}, where time corresponds to the date of MMSE score or prescribed medication as presented in Table 3. For the "Sig" model, the input variable were first transformed into signatures, where the categorical medication names were one-hot encoded and augmented with numerical MMSE scores to create a path. For the "Non-sig" model, the longitudinal MMSE scores were summarised by means of linear models adjusting for each patients and the median number of distinct medications were computed. Both models were trained and validated using the same folds of stratified 5-fold cross validation (with fixed random seed). The quality of predictions was assessed using the Harell's C-index and the results are summarised in Table 7. The signatures were computed using the "esig" Python library[3], however, alternative libraries are also available (Reizenstein and Graham, 2018; Kidger and Lyons, 2020).

Features	Sig	Non-sig
{time, MMSE}	0.626(0.009)	0.574(0.022)
{time, MMSE, meds}	0.621(0.011)	0.571(0.019)

Table 7: Harell's C-index measure of four models. Values reported as mean(std) over 5-fold cross validation.

We also estimated the time-dependent area under the curve of receiver operating characteristics (Lambert and Chevret, 2016). It is a natural extension of a common AUC ROC analysis to possibly censored survival times where the patients' cognitive health is usually better at the very first visit to a memory clinic, while their condition may deteriorate later. The time-dependent cumulative dynamic AUC ROC of all four models are presented in Fig. 2. The signature features outperformed the non-signature ones at all times and the inclusion of sequential information from switching medications improved AUC ROC at later times. However, both models struggle to reliably predict the future outcomes further than 3 years. This is due to the limitation of predictors and the available number of patients after 3 years rather than the capacity of our model.

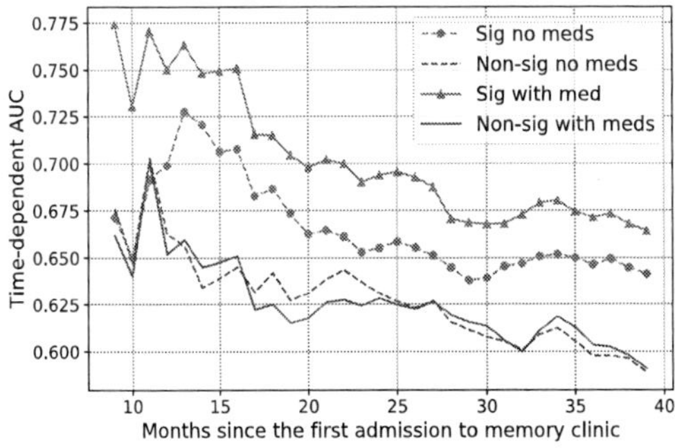

Figure 2: Time-dependent AUC of risk prediction over time since the first admission to a memory clinic.

4 Discussion and future direction

Unstructured longitudinal electronic health records, such as free-text clinical notes, inherently contain rich information about patients' health and outcomes over time. The right analytical tools capable of capturing sequential information can therefore maximise utilisation of longitudinal EHRs and can be valuable for supporting clinical decisions and prognostic models. In this work we implemented a signature-based approach to represent chronological events extracted using natural language processing from clinical notes. Extracted chronological events can be seen as a trajectory (path) embedded in a high-dimensional multi-modal space of events (i.e. different medications, interventions, measures, etc) and the signature uniquely characterises the path in the most succinct way. The signature-based feature extraction approach was compared to hand-crafted features, comprising a slope and an intercept of MMSE scores over time and the median number of medications for each patient. The signatures represent a hierarchical collection of features, where the first order is proportional to linear statistical moments (i.e. mean) and is not sensitive to the order of data points, as illustrated in Table 1. We demonstrated that the sequential information about medications has significantly improved the time-demented AUC as captured by the signatures (Figure 2). In future works we will extend the proposed framework to include the structured information available in EHR (i.e. lab results, coded procedures or clinical encounters) and will develop an interpretability framework to make the signature-based models explainable.

Acknowledgments

The study was funded by the National Institute for Health Research's (NIHR) Oxford Health Biomedical Research Centre (BRC-1215-20005). This work was supported by the UK Clinical Records Interactive Search (UK-CRIS) system funded and developed by the NIHR Oxford Health BRC at Oxford Health NHS Foundation Trust and the Department of Psychiatry, University of Oxford. AK, NV, QL, ANH are funded by the MRC Pathfinder Grant (MC-PC-17215). HN was supported by the Alan Turing Institute under the EPSRC grant EP/N510129/1 and under the EPSRC EP/S026347/1. The views expressed are those of the authors and not necessarily those of the UK National Health Service, the NIHR, or the UK Department of Health. We highly appreciate the work and support of the Oxford CRIS Team: Tanya Smith, Lulu Kane, Adam Pill and Suzanne Fisher and Dr Peter Phiri of the Southern Health NHS Foundation Trust CRIS team.

---

[3]https://esig.readthedocs.io/

References

Imanol Perez Arribas. 2018. Derivatives pricing using signature payoffs. *arXiv preprint arXiv:1809.09466*.

Imanol Perez Arribas, Kate Saunders, Guy Goodwin, and Terry Lyons. 2018. A signature-based machine learning model for bipolar disorder and borderline personality disorder. *Translational Psychiatry*, 8.

Dzmitry Bahdanau, Kyunghyun Cho, and Yoshua Bengio. 2014. Neural machine translation by jointly learning to align and translate. *arXiv preprint arXiv:1409.0473*.

Tian Bai, Shanshan Zhang, Brian L Egleston, and Slobodan Vucetic. 2018. Interpretable representation learning for healthcare via capturing disease progression through time. In *Proceedings of the 24th ACM SIGKDD International Conference on Knowledge Discovery & Data Mining*, pages 43–51.

Ilya Chevyrev and Andrey Kormilitzin. 2016. A primer on the signature method in machine learning. *arXiv preprint arXiv:1603.03788*.

Ilya Chevyrev, Vidit Nanda, and Harald Oberhauser. 2018. Persistence paths and signature features in topological data analysis. *IEEE transactions on pattern analysis and machine intelligence*, pages 1–1.

Ilya Chevyrev and Harald Oberhauser. 2018. Signature moments to characterize laws of stochastic processes. *arXiv preprint arXiv:1810.10971*.

David Collett. 2015. *Modelling survival data in medical research*. CRC press.

Joint Formulary Committee et al. 2019. *BNF 77 (British National Formulary) March 2019*. Pharmaceutical Press.

SM Goodday, A Kormilitzin, N Vaci, Q Liu, A Cipriani, T Smith, and A Nevado-Holgado. 2020. Maximizing the use of social and behavioural information from secondary care mental health electronic health records. *Journal of Biomedical Informatics*, page 103429.

Ben Hambly and Terry Lyons. 2010. Uniqueness for the signature of a path of bounded variation and the reduced path group. *Annals of Mathematics*, pages 109–167.

Henk Harkema, John N Dowling, Tyler Thornblade, and Wendy W Chapman. 2009. Context: an algorithm for determining negation, experiencer, and temporal status from clinical reports. *Journal of biomedical informatics*, 42(5):839–851.

Frank E Harrell, Robert M Califf, David B Pryor, Kerry L Lee, and Robert A Rosati. 1982. Evaluating the yield of medical tests. *Jama*, 247(18):2543–2546.

Sam Henry, Kevin Buchan, Michele Filannino, Amber Stubbs, and Ozlem Uzuner. 2020. 2018 n2c2 shared task on adverse drug events and medication extraction in electronic health records. *Journal of the American Medical Informatics Association*, 27(1):3–12.

Hemant Ishwaran, Udaya B Kogalur, Eugene H Blackstone, Michael S Lauer, et al. 2008. Random survival forests. *The annals of applied statistics*, 2(3):841–860.

Alistair EW Johnson, Tom J Pollard, Lu Shen, H Lehman Li-wei, Mengling Feng, Mohammad Ghassemi, Benjamin Moody, Peter Szolovits, Leo Anthony Celi, and Roger G Mark. 2016. Mimic-iii, a freely accessible critical care database. *Scientific data*, 3:160035.

Patrick Kidger, Patric Bonnier, Imanol Perez Arribas, Cristopher Salvi, and Terry Lyons. 2019. Deep signature transforms. In *Advances in Neural Information Processing Systems*, pages 3105–3115.

Patrick Kidger and Terry Lyons. 2020. Signatory: differentiable computations of the signature and logsignature transforms, on both cpu and gpu. *arXiv preprint arXiv:2001.00706*.

AB Kormilitzin, KEA Saunders, PJ Harrison, JR Geddes, and TJ Lyons. 2016. Application of the signature method to pattern recognition in the cequel clinical trial. *arXiv preprint arXiv:1606.02074*.

Andrey Kormilitzin, Kate EA Saunders, Paul J Harrison, John R Geddes, and Terry Lyons. 2017. Detecting early signs of depressive and manic episodes in patients with bipolar disorder using the signature-based model. *arXiv preprint arXiv:1708.01206*.

Andrey Kormilitzin, Nemanja Vaci, Qiang Liu, and Alejo Nevado-Holgado. 2020. Med7: a transferable clinical natural language processing model for electronic health records. *arXiv preprint arXiv:2003.01271*.

Jérôme Lambert and Sylvie Chevret. 2016. Summary measure of discrimination in survival models based on cumulative/dynamic time-dependent roc curves. *Statistical methods in medical research*, 25(5):2088–2102.

Guillaume Lample, Miguel Ballesteros, Sandeep Subramanian, Kazuya Kawakami, and Chris Dyer. 2016. Neural architectures for named entity recognition. *arXiv preprint arXiv:1603.01360*.

Shujian Liao, Terry Lyons, Weixin Yang, and Hao Ni. 2019. Learning stochastic differential equations using rnn with log signature features. *arXiv preprint arXiv:1908.08286*.

Terry Lyons. 2014. Rough paths, signatures and the modelling of functions on streams. *arXiv preprint arXiv:1405.4537*.

Fen Miao, Yun-Peng Cai, Yuan-Ting Zhang, and Chun-Yue Li. 2015. Is random survival forest an alternative to cox proportional model on predicting cardiovascular disease? In *6TH European conference of the international federation for medical and biological engineering*, pages 740–743. Springer.

James Morrill, Patrick Kidger, Cristopher Salvi, James Foster, and Terry Lyons. 2020a. Neural cdes for long time series via the log-ode method. *arXiv preprint arXiv:2009.08295*.

James Morrill, Andrey Kormilitzin, Alejo Nevado-Holgado, Sumanth Swaminathan, Sam Howison, and Terry Lyons. 2019. The signature-based model for early detection of sepsis from electronic health records in the intensive care unit. In *2019 Computing in Cardiology (CinC)*, pages Page–1. IEEE.

James H Morrill, Andrey Kormilitzin, Alejo J Nevado-Holgado, Sumanth Swaminathan, Samuel D Howison, and Terry J Lyons. 2020b. Utilization of the signature method to identify the early onset of sepsis from multivariate physiological time series in critical care monitoring. *Critical Care Medicine*, 48(10):e976–e981.

Verna C Pangman, Jeff Sloan, and Lorna Guse. 2000. An examination of psychometric properties of the mini-mental state examination and the standardized mini-mental state examination: implications for clinical practice. *Applied Nursing Research*, 13(4):209–213.

Jeffrey Pennington, Richard Socher, and Christopher D Manning. 2014. Glove: Global vectors for word representation. In *Proceedings of the 2014 conference on empirical methods in natural language processing (EMNLP)*, pages 1532–1543.

Sebastian Pölsterl, Nassir Navab, and Amin Katouzian. 2015. Fast training of support vector machines for survival analysis. In *Joint European Conference on Machine Learning and Knowledge Discovery in Databases*, pages 243–259. Springer.

James Pustejovsky and Amber Stubbs. 2012. *Natural Language Annotation for Machine Learning: A guide to corpus-building for applications.* " O'Reilly Media, Inc.".

Jeremy Reizenstein and Benjamin Graham. 2018. The iisignature library: efficient calculation of iterated-integral signatures and log signatures. *arXiv preprint arXiv:1802.08252*.

Isabel Segura Bedmar, Paloma Martínez, and María Herrero Zazo. 2013. Semeval-2013 task 9: Extraction of drug-drug interactions from biomedical texts (ddiextraction 2013). Association for Computational Linguistics.

Nemanja Vaci, Ivan Koychev, Chi-Hun Kim, Andrey Kormilitzin, Qiang Liu, Christopher Lucas,

Azad Dehghan, Goran Nenadic, and Alejo Nevado-Holgado. 2020. Real-world effectiveness, its predictors and onset of action of cholinesterase inhibitors and memantine in dementia: retrospective health record study. *The British Journal of Psychiatry*, page 1–7.

Zecheng Xie, Zenghui Sun, Lianwen Jin, Hao Ni, and Terry Lyons. 2017. Learning spatial-semantic context with fully convolutional recurrent network for online handwritten chinese text recognition. *IEEE transactions on pattern analysis and machine intelligence*, 40(8):1903–1917.

Weixin Yang, Terry Lyons, Hao Ni, Cordelia Schmid, Lianwen Jin, and Jiawei Chang. 2017. Leveraging the path signature for skeleton-based human action recognition. *arXiv preprint arXiv:1707.03993*.

Dongyu Zhang, Jidapa Thadajarassiri, Cansu Sen, and Elke Rundensteiner. 2020. Time-aware transformer-based network for clinical notes series prediction. In *Machine Learning for Healthcare Conference*, pages 566–588. PMLR.

Defining and Learning Refined Temporal Relations
in the Clinical Narrative

Chen Lin[1] (co-first author), Kristin Wright-Bettner[2] (co-first author),
Timothy Miller[1], Steven Bethard[3], Dmitriy Dligach[4],
Martha Palmer[2], James H. Martin[2] and Guergana Savova[1]

[1]Boston Children's Hospital and Harvard Medical School, Boston, MA
[2]Departments of Linguistics and Computer Science, University of Colorado, Boulder, CO
[3]School of Information, University of Arizona, Tucson, AZ
[4]Department of Computer Science, Loyola University Chicago, Chicago, IL
[1]{first.last}@childrens.harvard.edu
[2]{first.last}@colorado.edu
[3]bethard@arizona.edu
[4]ddligach@luc.edu

Abstract

We present refinements over existing temporal relation annotations in the Electronic Medical Record clinical narrative. We refined the THYME corpus annotations to more faithfully represent nuanced temporality and nuanced temporal-coreferential relations. The main contributions are in re-defining CONTAINS and OVERLAP relations into CONTAINS, CONTAINS-SUBEVENT, OVERLAP and NOTED-ON. We demonstrate that these refinements lead to substantial gains in learnability for state-of-the-art transformer models as compared to previously reported results on the original THYME corpus. We thus establish a baseline for the automatic extraction of these refined temporal relations. Although our study is done on clinical narrative, we believe it addresses far-reaching challenges that are corpus- and domain- agnostic.

1   Introduction

Temporal relation extraction and reasoning in the clinical domain continues to be a primary area of interest due to the potential impact on disease understanding and, ultimately, patient care. A significant body of text available for this purpose is the THYME (Temporal Histories of Your Medical Events) corpus (Styler IV et al., 2014), consisting of 594 clinical and pathology notes on colon cancer patients and 600 radiology, oncology and clinical notes on brain cancer patients, all from the Electronic Medical Record (EMR) of a

leading US medical center. This dataset has previously undergone a variety of annotation efforts, most notably temporal annotation (Styler IV et al., 2014). It has been part of several SemEval shared tasks such as Clinical TempEval (Bethard et al., 2017) where state-of-the-art results have been established. Our goal was to utilize this THYME corpus to enable the extraction of more extensive patient timelines by manually creating cross-document links that built off the pre-existing single file annotations.

(Wright-Bettner et al., 2019) discuss that a subset of the THYME temporal annotations contributed to incompatible temporal inferences, thus reducing their ability to support meaningful temporal reasoning. Accuracy and informativeness of temporal relation gold annotations are essential for their effectiveness in training a system for temporal relation extraction. We build on this work by offering an in-depth discussion of three key temporal relations − CONTAINS, CONTAINS-SUBEVENT (abbreviated as CON-SUB), and NOTED-ON − and how the addition of the last two types enhances the learnability of the temporal relations by resolving the conflicting temporal information in the original annotations.

While these revisions are corpus-specific, the reasoning behind them has far-reaching implications for automated timeline extraction. Since the cross-document linking task inherently deals with multiple, discrete narratives, it exposes the practical impact of discourse context on word

Proceedings of the 11th International Workshop on Health Text Mining and Information Analysis, pages 104–114
November 20, 2020. ©2020 Association for Computational Linguistics
https://doi.org/10.18653/v1/P17

sense (different narratives have different goals, which in turn influences meaning interpretation). This is discussed in detail in Section 4. We empirically found it essential to take changes in discourse context into account and suggest the same would be true for any annotation project that is interested in temporal reasoning, particularly those dealing with longer timelines (i.e., beyond the single-document level).

Recent developments in natural language processing establish neural approaches and more specifically transformer-based methods as the state of the art. Pre-trained models such as BERT (Devlin et al., 2019), BioBERT (Lee et al., 2020), Xlnet (Yang et al., 2019), ALBERT (Lan et al., 2020), RoBERTa (Liu et al., 2019), BART (Lewis et al., 2019), and SpanBERT (Joshi et al., 2020) report significant gains on multiple tasks. Thus, we demonstrate the learnability of the refined temporal relations in the context of these recent methodological developments.

2   Dataset

The 594 notes that make up the colon cancer part of the THYME corpus are grouped into sets, each set pertaining to a single patient and consisting of three notes written at different times during the patient's course of care. These notes had been previously annotated for five different intra-document temporal relations (BEFORE, OVERLAP, BEGINS-ON, ENDS-ON and CONTAINS), a subset of the ISO-TimeML temporal link (TLINK) types (Pustejovsky et al., 2010, Styler IV et al., 2014)[1]. To keep annotation manageable and circumvent massively inferential temporal linking, the THYME guidelines constrained TLINK creation to events within the same sentence or adjacent sentences, and specifically prohibited TLINKing across sections (these are clinically-delineated sections separated from each other by numerical section IDs – *History of Present Illness* is section 20103, *Vital Signs* is section 20110, etc.) Linguistic evidence for creating these TLINKs included local cues such as temporal

prepositions and adjectives (e.g., *during, subsequent to, prior to*), chronological narrative progression, and so forth.

Additionally, the notes had been separately annotated for intra-document coreference (IDENTICAL) and bridging (SET-SUBSET, WHOLE-PART) relations, which were later merged with the temporal annotations (Wright-Bettner et al., 2019). Temporal relations alone are insufficient for timeline extraction; coreference relations are also necessary (is the tumor seen in September the same as the one seen in March?).

Pursuant to our goal of reasoning over large-scale timelines, we built off these pre-existing within-note annotations by manually adding coreference and bridging links across each set of three notes. In the process, we discovered a subset of the original CONTAINS relations contributed to temporally-conflicting information, which led to the addition of two new TLINKs: NOTED-ON and CON-SUB (Wright-Bettner et al., 2019). We discuss below how these updates contribute to more accurate and comprehensive temporal relations which facilitated cross-document linking. As such, this is one of the few studies in clinical NLP for cross-document temporal relation annotations (see Raghavan et al., 2014 and Wright-Bettner et al., 2019; also see Song et al., 2018 for general domain cross-document temporal annotation discussions).

3   Refined Temporal Relation: NOTED-ON

The THYME guidelines specified that tests should always be in a CONTAINS relation with their results, as in (1), observing that it is inferential to say that something seen on the test exists before or after the test.

> 1. *January 17, 2009:* **CT** *abdomen and pelvis shows liver* **metastases**.
>   a.   *CT* CONTAINS *metastases*

We agree with the authors in part, particularly for a project that did not have coreference relations and that relied heavily on explicitly-temporal cues (*after, during, before*, etc.).

However, once the coreference links had been merged with the temporal annotations, more information was revealed about the temporal nature of the events. For example, *metastases* in

---

[1] This approach to temporal annotation in fact evolved directly from the ISO-TimeML model, in collaboration with Prof. James Pustejovsky. The Basic Formal Ontology has introduced similar relations (Smith et al., 2005), and a post-hoc mapping could be done between the relations proposed here and those in BFO 2.0. However, we defer to more philosophical scholars to determine the similarities and differences between ISO TimeML and BFO 2.0.

(1) might well be IDENTICAL to a later mention, such as:

> 2. *February 20, 2009, liver **metastases** resected with clear margins.*

The merged THYME and coreference annotations[2] for (1) and (2) were as follows:

- *January 17, 2009* CONTAINS *CT*
- *CT* CONTAINS *metastases*
- *February 20, 2009* CONTAINS *resected*
- *metastases* OVERLAPS *resected*
- *metastases* IDENTICAL *metastases*

Together, these relations entail that the same liver abnormalities were temporally contained by January 17, 2009 and temporally overlapped February 20, 2009 – a logical impossibility. This situation was extremely frequent in the data, reducing the informativeness of the CONTAINS links and the timeline as a whole. We therefore manually converted these links to a subtype of OVERLAP relation, NOTED-ON. NOTED-ON conveys temporal overlap between events and additionally says that Event A (*result*) was observed on/by Event B (*test*). (1a) was therefore re-annotated as follows:

> 3. *metastases* NOTED-ON *CT*

The application of this link was so constrained that it was fairly easy to annotate; we therefore added it as a single-annotated post-processing step.

4 Refined Temporal Relation: CONTAINS-SUBEVENT

4.1 Motivation and annotation

Early in the cross-document annotation pilot, we found the pre-existing intra-document schema categories were insufficient for dealing with cross-narrative phenomena, which in turn led to the addition of the new CON-SUB relation. Consider:

> 4. **Note A**: 2009: *Patient presents with recurrent **adenocarcinoma**. CT abdomen and pelvis ordered for further staging of this colon **cancer**.*
> > i. *adenocarcinoma* IDENT *cancer*

> **Note B**: *Successful removal of primary **tumor** in 2004. Recurrent **adenocarcinoma** is inoperable. Patient here today to discuss other treatment options for his colon **cancer**.*
> > ii. *cancer* CONTAINS *tumor*
> > iii. *cancer* CONTAINS *adenocarcinoma*

Both sets of intra-document links are pragmatically appropriate. Discourse contexts can expand or reduce the level of granularity at which a sense is interpreted (Recasens et al., 2011; also see Hobbs, 1985). In Note A, the text supports a coarse-grained interpretation of *adenocarcinoma*, or what Hovy et al., 2013 term a "wide" reading; it refers generally to the patient's cancer. Note B, however, requires a fine-grained ("narrow") interpretation – *adenocarcinoma* here refers specifically to the new, inoperable tumor and is contrasted with the original, resected tumor.

The quandary for the cross-document task lies in whether to link *adenocarcinoma* in A as IDENTICAL to *adenocarcinoma* in B. An IDENTICAL relation entails logical impossibilities: assuming we also link *cancer*ₐ as IDENT to *cancer*ᵦ, the combined within- and cross-document relations now say the recurrent adenocarcinoma temporally contains itself *and* the primary tumor which was removed years earlier. This reduces the meaningfulness of the CONTAINS links and therefore the timeline. On the other hand, leaving the two *adenocarcinoma* references unlinked fails to capture the significant semantic relation between two identical strings (*recurrent adenocarcinoma*) that do in fact refer, on some level of granularity, to the same real-world event. In either case, annotators are stymied.

Clearly, the pre-existing intra-document schema categories, specifically the binary coreference choice ($A = B$ or $A \neq B$), were insufficient for dealing with the sense variation and nuanced event structure exposed by multiple narratives. We therefore introduced CON-SUB (based on O'Gorman et al., 2016) as a subtype of the CONTAINS TLINK type. While CONTAINS conveys only temporal containment, CON-SUB *additionally* says that Event B is intrinsically part

---

of the structure of Event A[3]. The difference may be seen in (5):

5. *During patient's neoadjuvant **treatment**, she was in a car **accident** which **delayed cycle 4**.*
    a. *treatment* CONTAINS *accident, delayed*
    b. *treatment* CON-SUB *cycle*

We were then able to re-annotate the intra-document relations for *both* notes A and B above as *cancer* CON-SUB *adenocarcinoma*, which streamlined cross-document decisions – *cancer$_A$ IDENT cancer$_B$* and *adenocarcinoma$_A$ IDENT adenocarcinoma$_B$* – while preserving the "quasi-identical" relation (Hovy et al., 2013) between the cancer and adenocarcinoma terms. While this solution does not fully resolve the problem of binary annotation choices, it does provide more "wiggle room" along the meaning spectrum (Cruse, 1986) by introducing a third value – two mentions may be identical, non-identical, or mereologically (part-whole) related.

In keeping with our primary goal of enabling timeline extraction, we implemented CON-SUB as a TLINK since it conveys true temporal containment. CON-SUB, however, differs from other TLINKs, which were constrained by proximity and lexical cues, as discussed in section 2. The fact that CON-SUB also represents structural information allowed us to treat it like a coreference/bridging relation in terms of permissible textual evidence for link creation: namely, semantic scripts. These may be defined as: "a stereotypical sequence of events" (Araki et al., 2014) or "prototypical schematic sequences of events" (Chambers and Jurafsky, 2008). We can expect a surgery, for example, to consist of certain, typical subevents (incisions, subprocedures, anesthesia administration, etc.), which therefore enables annotators to look throughout the whole document for lexical items with meanings that fit those subevents. The concept of semantic scripts is what facilitates attainable long-distance coreference /bridging linking, and therefore, long-distance CON-SUB linking. This is obviously not the case for non-subevent CONTAINS relations.

We were therefore able to revise the THYME annotations to accommodate CON-SUB in two ways: First, we converted CONTAINS links to CON-SUB as appropriate, e.g.: *treatment CONTAINS radiation* became *treatment CON-SUB radiation*. Secondly, we added certain long-distance CON-SUB links for which there were no pre-existing CONTAINS annotations. These changes were made via a double-blind annotation process followed by an adjudication pass. The annotation team for the entire project (including the cross-document stage) consisted of nine annotators, eight of whom either had or were obtaining undergraduate or graduate degrees in linguistics. The ninth annotator was a physician who received on-the-job linguistics training and focused primarily on annotation subtasks that demanded considerable medical knowledge. Additionally, we consulted regularly with an oncologist and a medical coder with a decade of NLP annotation experience.

4.2   Inter-annotator Agreement

While the gold intra-document CON-SUB relations enabled high cross-document coreference agreement (93.77%), the IAA score for single-file CON-SUB links themselves was low at 34.14%[4]. Several factors contributed to this, but we focus on one major one here. The size and complexity of the guidelines[5] reflected the size and complexity of the task, augmenting the already heavy cognitive burden on annotators.

The cross-document task exposes greater nuance in event structure, involves greater variability of word sense, and attempts to join narratives that are temporally and linguistically disjunct. Annotation guidelines that set out to accurately represent information that is inherently nuanced, variable, and noncohesive will not be simple (see Savkov et al., 2016).

In determining guidelines for the subevent relation, we found that the best course for handling variability in lexical sense differed depending on the semantic potential (that is, how

---

[3] We did not use the pre-existing WHOLE-PART relation in order to preserve a distinction between events and entities, which also streamlined the annotation process. (WHOLE-PART was most often used to represent anatomical-site relations, such as *colon$_{WHOLE}$ – splenix flexure$_{PART}$*.)

[4] This score represents annotator-annotator agreement. We did not score annotator-gold, since adjudicators were permitted to make some specific changes that annotators were not.

[5] https://www.colorado.edu/lab/clear/projects/computationalsemantics/annotation

adaptable a word is to different meanings; see Evans, 2006) of the individual words used most frequently for each event category (i.e., semantic script)[6]. Not surprisingly, lexically-specific rules contributed significantly to the sheer size and complexity of the guidelines. Compare the following:

> 6. *We are seeing the patient for recent diagnosis of colon **cancer**. The **tumor** in her colon is quite large.*

> 7. *We recommended adjuvant **treatment**. Patient will return to start **chemo** in two weeks.*

Unmodified, *treatment* has an impoverished semantic potential (Evans, 2006); the meaning it conveys in itself is sparse, yielding a semantic flexibility that allows it to represent a wide range of referents. It is intuitive to understand it in (7) as coreferential with the much more precise *chemo*. *Cancer*, however, has a richer potential; it conveys a temporally-extensive disease that may have multiple manifestations (a primary tumor, recurrent tumors, metastatic tumors, etc.), rendering it less amenable to a coreferential link with *tumor*.

Therefore, for the cancer semantic script, annotators were asked to distinguish between terms that are defined more generally (*cancer, disease*) and terms that are more specific (*adenocarcinoma, tumor, metastasis, mass*, etc.), such that the specific terms were always subevents of the general terms, *regardless of pragmatic support for wide or narrow readings for a given term*. The reason for this has already been partially discussed in example (4); we add here that the semantic rigidity of *cancer* (compared to *treatment/therapy* terms, for example) also informed this choice.

This required a degree of abstraction from the text and an intentional suppression of instinctive linguistic judgments on the single-document level – for example, *adenocarcinoma* in (4) was re-annotated as a subevent of *cancer*, in spite of the fact that the text supports the wide reading.[7]

On the other hand, for the treatment/therapy semantic script, annotators were to determine relations more intuitively. Compare the resulting impact for cross-document linking in (8) to (4):

> 8. **Note A**: *We recommended adjuvant **treatment**. Patient will return for the first day of **chemo** in two weeks.*
>   a. *treatment* IDENT *chemo*
> **Note B**: *Adjuvant **treatment** consisted of four months of **chemo** and **radiation** and was without complication.*
>   b. *treatment* CON-SUB *chemo*
>   c. *treatment* CON-SUB *radiation*

If we linked *treatment$_A$* to *treatment$_B$* and *chemo$_A$* to *chemo$_B$*, the product would be the same undesirable entailments discussed in (4): in this case, that radiation is a subevent of chemotherapy (an entirely different treatment), and that the chemo event temporally contains itself. Here, however, our solution differed: Rather than re-annotate (8a) as *treatment* CON-SUB *chemo*, we simply chose to leave *treatment$_A$* and *treatment$_B$* unlinked in cross-document annotation. The only cross-document link for this context was *chemo$_A$* IDENT *chemo$_B$*. Again, this is due to the semantic malleability of *treatment*. Leaving two identical strings (*adjuvant treatment*) unlinked to each other is less problematic when that string regularly refers to a wide variety of referents. Furthermore, in experiments with re-analysis that paralleled the cancer semantic script approach, we found that attempting to always annotate *treatment* as an umbrella event proliferated unnecessary nested relations (because of how modifiable it is), increasing disagreement potential.

Unsurprisingly, an analysis of CON-SUB disagreements suggests that annotators struggled to remember when to abstract terms away from the context and when to interpret them intuitively; an example like (7) was apt to produce a disagreement, shown here in (9):

---

[6] Due to time constraints, we only added subevent relations for four categories of events: cancer, cancer treatment (surgeries, chemotherapy, and radiation), medications, and chronic diseases.

[7] In and of itself, this is not unreasonable given that the cross-document annotation task inherently lacks the linguistic cues supplied by a single, cohesive discourse (Wright-Bettner et al., 2019).

Relation type	Description	Link
CONTAINS supertype		
CONTAINS	M1 temporally contains M2	[M1] CONTAINS [M2]
CONTAINS-SUBEVENT	M1 temporally contains M2 and M2 is part of the structure of M1	[M1] CON-SUB [M2]
OVERLAP supertype		
OVERLAP	M1 temporally overlaps M2	[M1] OVERLAPS [M2]
NOTED-ON	M1 temporally overlaps and is observed on M2 (used to link tests to test results)	[M1] NOTED-ON [M2]
Other temporal links		
BEFORE	M1 temporally begins and ends before M2 begins	[M1] BEFORE [M2]
BEGINS-ON	The start of M1 begins at the end of M2	[M1] BEGINS-ON [M2]
ENDS-ON	The end of M1 ends at the start of M2	[M1] ENDS-ON [M2]

Table 1: Revised intra-document temporal links in the THYME colon cancer corpus

9. We recommended adjuvant **treatment***. Patient will return to start* **chemo** *in two weeks.*

> *Annotator A: treatment IDENT chemo*
> *Annotator B: treatment CON-SUB chemo*

While Annotator A correctly interpreted the terms as coreferential (based on the context), Annotator B mistakenly followed the approach for the cancer semantic script in analyzing "treatment" as an umbrella event.

The annotation task was already demanding, requiring annotators to learn and assimilate specialized medical knowledge and terminology from the clinical texts, which themselves are written in heavy shorthand and with a mix of template language and free text that sometimes conflict. In addition, the non-linguistic nature of the cross-document component forced the creation of several counterintuitive annotation rules, which frequently (but not always) required them to ignore real linguistic cues.

Finally, we did not calculate IAA for non-subevent CONTAINS links because they already existed. They did, however, change slightly, along with all the TLINK types, since the auto-merger of the temporal and coreference annotations (discussed in section 2) produced some informational conflicts that we calibrated in a manual single-annotated pre-processing pass. However, as a proxy for annotator agreement, we show below that learnability improved for all temporal relations.

5 Summary of Refined THYME Relations

The pre-existing CONTAINS annotations were revised in part through the addition of two new links, both of which convey temporal and non-temporal information. The original CONTAINS and OVERLAP relations were therefore reimagined as supertypes[8], each consisting of two subtypes. All links are described in Table 1. We refer to these fine-grained THYME annotations as THYME+ and to the original THYME annotations as THYME.

In the next section we demonstrate the learnability of the refined THYME+ temporal relations with state-of-the-art transformer methods. We report results that establish baselines for the THYME+ corpus for further methods development and offer insights into the challenges we faced which we view as exciting venues for future research.

6 Learning Refined Temporal Relations

Following the same window-based processing (using a span of contiguous tokens disregarding sentence boundaries for generating relational candidates) and argument-marking mechanism developed by the prior study that achieved the state-of–the-art results on THYME (Lin et al.,

---

[8] It is worth noting that while CONTAINS itself may be thought of as a specific subtype of an overlap temporal relation, we use the OVERLAP category specifically for non-containment temporal overlap or underspecified cases for which we cannot claim containment.

2019), we tested a series of pre-trained models for extracting both within and cross-sentence temporal relations (i.e. TLINKs) in a multi-class classification fashion. Figure 1 shows a CON-SUB relation between "cancer" and "adenocarcinoma", and its representation as a token sequence. Special token pairs, "eas" and "eae", "ebs" and "ebe" mark the events of interest in the sequence.

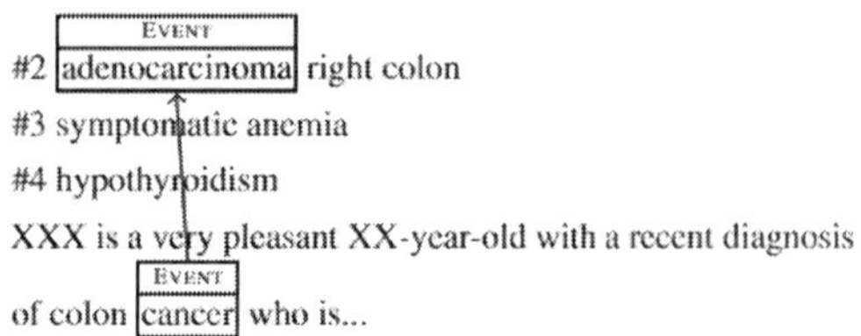

The above CON-SUB instance is represented as ⇓
2 **eas** adenocarcinoma **eae** right colon newline # 3 symptomatic anemia newline # 4 hypothyroidism newline XXX is a very pleasant XX - year - old with a recent diagnosis of colon **ebs** cancer **ebe** who is

Figure 1: CON-SUB instance and its representation as a token sequence

Pre-trained models BERT (Devlin et al., 2019), BioBERT (Lee et al., 2020), Xlnet (Yang et al., 2019), ALBERT (Lan et al., 2020), RoBERTa (Liu et al., 2019), BART (Lewis et al., 2019), and SpanBERT (Joshi et al., 2020) were used to encode each input sequence with a relational candidate by the [CLS] token, which was fed to the classification layer to predict the relation type for every relational pair candidate. For some of the popular models, such as BERT, RoBERTa, and BART, we also tried their *large* version in addition to their *base* releases.

We used NVIDIA GTX Titan Xp GPU and Titan RTX GPU cluster of 7 nodes for fine-tuning the pre-trained models. The fine-tuning is done with HuggingFace's Transformers API (Wolf et al., 2019) and the TensorFlow-based BERT API, with batch size selected from (16, 32), a 60-token sliding window for generating candidate relational pairs, a maximal sequence length of 100 word pieces to accommodate all word pieces from the 60 tokens, and a learning rate selected from (1e-5, 2e-5, 3e-5, 5e-5). The performance was evaluated by the standard Clinical TempEval (Bethard et al., 2017) evaluation script, modified only to accommodate the new categories.

7 Experimental Results

The model that performed best on THYME (BioBERT-base) was trained and evaluated on THYME+ annotations. The first two rows of Table 2 gauge performance purely based on the refinements of the THYME+ annotations. Splitting CONTAINS into CONTAINS and CON-SUB relations and OVERLAP into OVERLAP and NOTED-ON relations leads to better learnability: CONTAINS goes from 0.664 F1 on THYME to 0.748 F1 on THYME+, and OVERLAP goes from 0.179 on THYME to 0.416 on THYME+. The best results for the new categories of CON-SUB and NOTED-ON are 0.072 F1 and 0.744 F1 respectively − results that establish baselines for these two new temporal relations. The performance on all types of relations for THYME+ is 0.625 F1 compared to 0.548 for THYME (Table 2, *Overall* column, rows 1 and 2).

Lin et al., 2019 report 0.684 F1 for THYME CONTAINS, however the result is achieved when training on and evaluating for only the CONTAINS links, and augmenting the training data with automatically generated CONTAINS relations. Thus, it is not a fair comparison to use for the results reported in Table 2.

Of the models beyond BioBERT that we explored, BART-large was the most successful. The result with BART-large was 0.748 F1 (Table 2, *CONTAINS* column, row 3). In general, certain pre-trained models, like BioBERT and BART, yield better results than the other models. BioBERT is pre-trained on biomedical text and thus can help encode clinical text better. BART masks a contiguous span of text rather than random tokens, which can be helpful for encoding clinical text where many event and temporal expressions consist of multiples tokens, e.g. "ascending colon cancer".

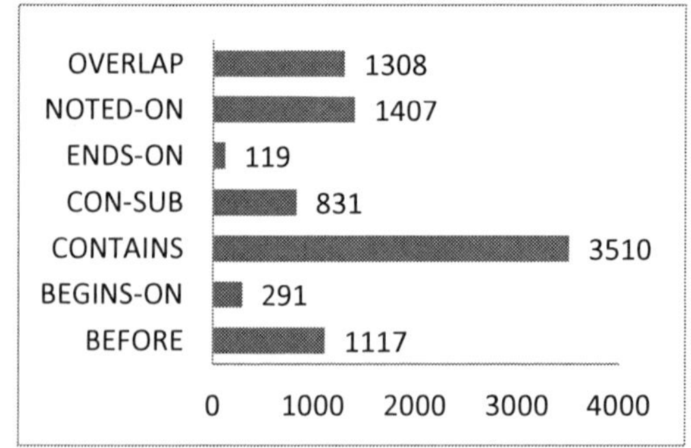

Figure 2: Number of instances for each temporal relation type in the colon test set.

Model	Data	BEFORE			BEGINS-ON			CONTAINS			CON-SUB		
		P	R	F1	P	R	F1	P	R	F1	P	R	F1
BioBERT	THYME	0.168	0.311	0.218	0.263	0.070	0.110	0.701	0.631	0.664			
BioBERT	THYME+	0.278	**0.458**	0.346	**0.423**	0.175	**0.248**	0.793	0.708	0.748			
BART	THYME+	**0.313**	0.434	**0.364**	0.383	**0.179**	0.244	0.791	0.709	0.748	0.375	0.040	0.072
BART-	THYME+	0.300	0.422	0.351	0.378	0.175	0.239	**0.796**	**0.710**	**0.750**			

Model	Data	ENDS-ON			NOTED-ON			OVERLAP			OVERALL		
		P	R	F1	P	R	F1	P	R	F1	P	R	F1
BioBERT	THYME	**0.194**	0.099	0.131				0.174	0.185	0.179	0.594	0.508	0.548
BioBERT	THYME+	0.112	**0.210**	0.146	**0.786**	0.706	**0.744**	0.353	**0.508**	0.416	0.696	**0.568**	0.625
BART	THYME+	0.120	**0.210**	0.153	0.787	0.702	0.742	0.401	0.482	**0.438**	0.713	0.567	**0.632**
BART-	THYME+	0.124	**0.210**	**0.156**	**0.786**	**0.707**	**0.744**	**0.404**	0.470	0.435	**0.718**	0.558	0.628

Table 2: Model performance on THYME and THYME+.
BART = BART large trained on all THYME+ links.
BART- = BART large trained on all THYME+ links excluding CON-SUB

CONTAINS and NOTED-ON are the two types of temporal relations that are overwhelmingly the most frequent in THYME+ (Figure 2), thus of the greatest interest from an applications point of view. THYME+ CONTAINS best result of 0.75 F1 and THYME+ NOTED-ON best result of 0.744 F1 are higher than the state-of-the-art THYME result on CONTAINS of 0.684 F1 (Lin et al., 2019). BEGINS-ON and ENDS-ON categories have the fewest instances, as shown in Figure 2, so their low performance could thus be due to their lack of representation. CON-SUB's low performance could be attributed to their coreferential nature and the long distance relations between the two arguments which in many cases surpass our 60-token window limit.

Table 2, row 4 presents the results with the BART-large model trained and evaluated when excluding CON-SUB links. The 0.750 F1 on CONTAINS is similar to the 0.748 F1 on CONTAINS when training and evaluating on all THYME+ relations. Thus, while the model is not able to accurately predict CON-SUB relations, including them does not appear to cause confusion for the model.

8 Discussion

Splitting CONTAINS into CONTAINS and CON-SUB categories improved the annotation quality of the CONTAINS class instances in THYME+. The CONTAINS class is the most frequent relation in clinical text and very easy for transitive closure to operate upon. As we already pointed out, the CONTAINS performance on THYME+ is improved from 0.664 F1 to 0.748 F1 using the same BioBERT model (Table 2, *CONTAINS* column, row 1 and 1), with both improved P and R.

The creation of gold NOTED-ON instances was straightforward, thus with high quality. NOTED-ON is the second most frequent relation in the corpus. 65.12% of the NOTED-ON relations are within one sentence and very few cases are long-distance. This makes the NOTED-ON class very learnable.

The results on THYME+ BEFORE, BEGINS-ON, ENDS-ON, and OVERLAP also improved compared to their respective THYME results. We attribute it to the improved performance of CONTAINS and NOTED-ON links as the definitions, hence the space separation, are tightened.

An error analysis of the CON-SUB relations showed that the main error consisted of missed links that relied on the semantic-script concept discussed in section 4.2. For example, the system often failed to capture the CON-SUB relation between *cancer* and *adenocarcinoma*. These are often long-distance relations: of all gold CON-SUB relations, 67.63% of them are beyond our 60-token window limit. Even if we focused on those within-window CON-SUB relations (32.37% of total), the performance was still low (0.441 P, 0.112 R, 0.178 F1), which showed our models had not captured the peculiarities of the CON-SUB class. The fact that the majority of instances of CON-SUB class are long-distance is quite different from the other TLINKs and hard for transitive closure to act upon, suggesting they might need a different approach than the other TLINKs.

However, one error category that could be resolved by transitive closure are links the system marked that are not present locally in the gold, but are correct by inference via the coreference relations. Consider:

10. ***Procedure*** to include hernia ***repair***.
a. System: *Procedure* CON-SUB *repair*

b. Gold: No local TLINK.

In this example, one or both terms are linked as IDENT to earlier mentions in the note. To save time and visual clutter, annotators only linked bridging relations like CON-SUB to first mentions, under the assumption that redundant information could be retrieved from the IDENT chains. Therefore, there is no error here if the IDENT chains are taken into account.

In short, CON-SUB relations capture temporality and mereological relations. Thus representations and methods for combined temporality and coreference are suitable venues to explore.

Except for CONTAINS, the other types of TLINKs have relatively low numbers of instances. The creation of instances for the low number of temporal relation types is limited by two main factors: (1) availability of data due to privacy constraints on EMR clinical narratives, and (2) the time, effort and budget required for such an activity. We have shown that with enough training instances (see CONTAINS), temporal relations are learnable at improved rates with the latest state-of-the-art methods. Although NOTED-ON has a similarly low number of instances as OVERLAP and BEFORE, it is highly learnable (0.744 F1) which we attribute to its semantic-script characteristics as discussed in section 4.2. This suggests that there are several paths to explore among which are: (1) re-defining and refining the other types of relations, and (2) devising methods for relations with low number of instances.

9   Conclusion

In this study, we presented our refinements for temporal relation annotations of the THYME corpus resulting in the THYME+ corpus. The main modifications are in re-defining CONTAINS and OVERLAP relations into CONTAINS, CONTAINS-SUBEVENT, OVERLAP and NOTED-ON. This strategy is theoretically based and led to better learnability with the latest transformer methods. Our results establish baselines for future methods -- CONTAINS 0.750 F1 OVERLAP 0.438 F1, CON-SUB 0.072 F1 and NOTED-ON 0.744 F1.

Acknowledgements

The work was supported by funding from the United States National Institutes of Health -- R01LM010090 from the National Library of Medicine and U24CA248010 and UG3CA243120 from the National Cancer Institute. The content is solely the responsibility of the authors and does not necessarily represent the official views of the National Institutes of Health.

We would like to thank: Samuel Beer, Hayley Coniglio, Jameson Ducey, Ahmed Elsayed, Dana Green, Matthew Oh, Michael Regan, and Adam Wiemerslage for annotation; Piet deGroen and David Harris for medical advice; and Wei-Te Chen and Skatje Myers for technical support.

References

Jun Araki, Zhengzhong Liu, Eduard Hovy, and Teruko Mitamura. 2014. Detecting Subevent Structure for Event Coreference Resolution. In *Proceedings of the Ninth International Conference on Language Resources and Evaluation (LREC'14)*, page 6, Reykjavik, Iceland, May.

Steven Bethard, Guergana Savova, Martha Palmer, and James Pustejovsky. 2017. SemEval-2017 Task 12: Clinical TempEval. In *Proceedings of the 11th International Workshop on Semantic Evaluation (SemEval-2017)*, pages 565–572, Vancouver, Canada, August. Association for Computational Linguistics.

Nathanael Chambers and Dan Jurafsky. 2008. Unsupervised Learning of Narrative Event Chains. In *Proceedings of ACL-08: HLT*, pages 789–797, Columbus, Ohio, June. Association for Computational Linguistics.

David Cruse. 1986. *Lexical Semantics*. Cambridge University Press, Cambridge, UK.

Jacob Devlin, Ming-Wei Chang, Kenton Lee, and Kristina Toutanova. 2019. BERT: Pre-training of Deep Bidirectional Transformers for Language Understanding. *arXiv:1810.04805 [cs]*, May. arXiv: 1810.04805.

Vyvyan Evans. 2006. Lexical concepts, cognitive models and meaning-construction. *Cognitive Linguistics*, 17(4):491–534.

Jerry R. Hobbs. 1985. Granularity. In *Proceedings of the Ninth International Joint Conference on Artificial Intelligence*, volume 1, pages 432–435. August.

Eduard Hovy, Teruko Mitamura, Felisa Verdejo, Jun Araki, and Andrew Philpot. 2013. Events are Not Simple: Identity, Non-Identity, and Quasi-Identity. In *Workshop on Events: Definition, Detection, Coreference, and Representation*, pages 21–28, Atlanta, Georgia, June. Association for Computational Linguistics.

Mandar Joshi, Danqi Chen, Yinhan Liu, Daniel S. Weld, Luke Zettlemoyer, and Omer Levy. 2020. SpanBERT: Improving Pre-training by Representing and Predicting Spans. *arXiv:1907.10529 [cs]*, January. arXiv: 1907.10529.

Zhenzhong Lan, Mingda Chen, Sebastian Goodman, Kevin Gimpel, Piyush Sharma, and Radu Soricut. 2020. ALBERT: A Lite BERT for Self-supervised Learning of Language Representations. *arXiv:1909.11942 [cs]*, February. arXiv: 1909.11942.

Jinhyuk Lee, Wonjin Yoon, Sungdong Kim, Donghyeon Kim, Sunkyu Kim, Chan Ho So, and Jaewoo Kang. 2020. BioBERT: a pre-trained biomedical language representation model for biomedical text mining. *Bioinformatics*, 36(4):1234–1240, February.

Mike Lewis, Yinhan Liu, Naman Goyal, Marjan Ghazvininejad, Abdelrahman Mohamed, Omer Levy, Ves Stoyanov, and Luke Zettlemoyer. 2019. BART: Denoising Sequence-to-Sequence Pre-training for Natural Language Generation, Translation, and Comprehension. *arXiv:1910.13461 [cs, stat]*, October. arXiv: 1910.13461.

Chen Lin, Timothy Miller, Dmitriy Dligach, Steven Bethard, and Guergana Savova. 2019. A BERT-based universal model for both within-and cross-sentence clinical temporal relation extraction. In *Proceedings of the 2nd Clinical Natural Language Processing Workshop*, pages 65–71, Minneapolis, MN.

Yinhan Liu, Myle Ott, Naman Goyal, Jingfei Du, Mandar Joshi, Danqi Chen, Omer Levy, Mike Lewis, Luke Zettlemoyer, and Veselin Stoyanov. 2019. RoBERTa: A Robustly Optimized BERT Pretraining Approach. *arXiv:1907.11692 [cs]*, July. arXiv: 1907.11692.

Tim O'Gorman, Kristin Wright-Bettner, and Martha Palmer. 2016. Richer Event Description: Integrating event coreference with temporal, causal and bridging annotation. In *Proceedings of the 2nd Workshop on Computing News Storylines (CNS 2016)*, pages 47–56, Austin, Texas, November. Association for Computational Linguistics.

James Pustejovsky, Kiyong Lee, Harry Bunt, and Laurent Romary. 2010. ISO-TimeML: An International Standard for Semantic Annotation. In *Proceedings of the Seventh International Conference on Language Resources and Evaluation (LREC'10)*, page 4, Valletta, Malta, May.

Preethi Raghavan, Eric Fosler-Lussier, Noémie Elhadad, and Albert M. Lai. 2014. Cross-narrative Temporal Ordering of Medical Events. In *Proceedings of the 52nd Annual Meeting of the Association for Computational Linguistics (Volume 1: Long Papers)*, pages 998–1008, Baltimore, Maryland, June. Association for Computational Linguistics.

Marta Recasens, Eduard Hovy, and M. Antònia Martí. 2011. Identity, non-identity, and near-identity: Addressing the complexity of coreference. *Lingua*, 121(6):1138–1152, May.

Aleksandar Savkov, John Carroll, Rob Koeling, and Jackie Cassell. 2016. Annotating patient clinical records with syntactic chunks and named entities: the Harvey Corpus. *Language Resources and Evaluation*, 50:523–548.

Barry Smith, Werner Ceusters, Bert Klagges, Jacob Köhler, Anand Kumar, Jane Lomax, Chris Mungall, Fabian Neuhaus, Alan L. Rector, and Cornelius Rosse. 2005. Relations in biomedical ontologies. *Genome Biology*, 6(5):R46, April.

Zhiyi Song, Ann Bies, Justin Mott, Xuansong Li, Stephanie Strassel, and Christopher Caruso. 2018. Cross-Document, Cross-Language Event Coreference Annotation Using Event Hoppers. In *Proceedings of the Eleventh International Conference on Language Resources and Evaluation (LREC 2018)*, Miyazaki, Japan, May. European Language Resources Association (ELRA).

William F. Styler IV, Steven Bethard, Sean Finan, Martha Palmer, Sameer Pradhan, Piet C de Groen, Brad Erickson, Timothy Miller, Chen Lin, Guergana Savova, and James Pustejovsky. 2014. Temporal Annotation in the Clinical Domain. *Transactions of the Association for Computational Linguistics*, 2:143–154.

Thomas Wolf, Lysandre Debut, Victor Sanh, Julien Chaumond, Clement Delangue, Anthony Moi, Pierric Cistac, Tim Rault, Rémi Louf, Morgan Funtowicz, Joe Davison, Sam Shleifer, Patrick von Platen, Clara Ma, Yacine Jernite, Julien Plu, Canwen Xu, Teven Le Scao, Sylvain Gugger, et al. 2019. HuggingFace's Transformers: State-of-the-art Natural Language Processing. *arXiv e-prints*, 1910:arXiv:1910.03771, October.

Kristin Wright-Bettner, Martha Palmer, Guergana Savova, Piet de Groen, and Timothy Miller. 2019. Cross-document coreference: An approach to capturing coreference without context. In *Proceedings of the Tenth International Workshop on Health Text Mining and Information Analysis (LOUHI 2019)*, pages 1–10, Hong Kong, November. Association for Computational Linguistics.

Zhilin Yang, Zihang Dai, Yiming Yang, Jaime Carbonell, Russ R Salakhutdinov, and Quoc V Le. 2019. XLNet: Generalized Autoregressive Pretraining for Language Understanding. In H. Wallach, H. Larochelle, A. Beygelzimer, F. d\textquotesingle Alché-Buc, E. Fox, and R. Garnett, editors, *Advances in Neural Information Processing Systems 32*, pages 5753–5763. Curran Associates, Inc.

Context-Aware Automatic Text Simplification of Health Materials in Low-Resource Domains

Tarek Sakakini, Jong Yoon Lee, Aditya Duri, Renato F.L. Azevedo, Kuangxiao Gu
University of Illinois at Urbana-Champaign
{sakakini, jlee642, duri2, razeved2, kgu3}@illinois.edu

Suma Bhat, Dan Morrow, Mark Hasegawa-Johnson, Thomas Huang
University of Illinois at Urbana-Champaign
{spbhat2, dgm, jhasegaw, t-huang1}@illinois.edu

Victor Sadauskas, James Graumlich, Saqib Walayat
University of Illinois College of Medicine, Peoria
{vsadau2, jfg, swalayat}@uic.edu

Ann Willemsen-Dunlap, and Donald Halpin
Jump Simulation Center, Peoria, Illinois
{ann.m.willemsen-dunlap, donald.j.halpin}@jumpsimulation.org

Abstract

Healthcare systems have increased patients' exposure to their own health materials to enhance patients' health levels, but this has been impeded by patients' lack of understanding of their health material. We address potential barriers to their comprehension by developing a context-aware text simplification system for health material. Given the scarcity of annotated parallel corpora in healthcare domains, we design our system to be independent of a parallel corpus, complementing the availability of data-driven neural methods when such corpora are available. Our system compensates for the lack of direct supervision using a biomedical lexical database: Unified Medical Language System (UMLS). Compared to a competitive prior approach that uses a tool for identifying biomedical concepts and a consumer-directed vocabulary list, we empirically show the enhanced accuracy of our system due to improved handling of ambiguous terms. We also show the enhanced accuracy of our system over directly-supervised neural methods in this low-resource setting. Finally, we show the direct impact of our system on laypeople's comprehension of health material via a human subjects' study ($n = 160$).

1 Introduction

Healthcare practices have granted patients increased access to their health information to support self-care (Davis et al., 2005; Detmer et al., 2008). But, the benefits have been hindered by patients' low comprehension of their own health data (Irizarry et al., 2015), as a study shows that readability measures of online health information is significantly higher than patient health literacy abilities (Mcinnes and Haglund, 2011). Moreover, older adults, the largest demographic group interacting with the healthcare system, are often the least health-literate (Kessels, 2003; Kutner et al., 2006). With low levels of health literacy resulting in worse health outcomes (Ha and Longnecker, 2010; Kindig et al., 2004), there is an urgent need to reduce the gap between the health literacy of patients and the health literacy demands of healthcare systems. Accordingly, we explore text simplification methods for health text, taking medication instructions as a use case (see Table 1).

This mismatch in patient literacy levels and health documents is due in part to the differing language used by healthcare professionals and patients (Rotegard et al., 2006). For example, what professionals refer to as "PO", patients might refer

Proceedings of the 11th International Workshop on Health Text Mining and Information Analysis, pages 115–126
November 20, 2020. ©2020 Association for Computational Linguistics
https://doi.org/10.18653/v1/P17

Original:
Take 50 mg PO daily for 2 days. Hold for SBP < 90.
Gold Simplification:
Take 50 milligrams by mouth daily for 2 days. Hold for systolic blood pressure < 90.
Dr. Babel Fish:
Take 50 milligrams orally daily for 2 days. Hold for systolic blood pressure < 90.

Table 1: Example medication instruction, its target simplification, and our system's simplification. Color coding reflects replacement.

to as "by mouth". While previous works have addressed this by performing local word replacement (Kandula et al., 2010), their context-free frameworks lacked the accuracy. In a health document, "Mg" could mean "milligrams" or "Magnesium", and harnessing the contextual information, for example in "Take 50 Mg" or "Mg reacts with", aids accurate simplification.

Our approach is a context-aware medical text simplification system, named Dr. Babel Fish (DBF). We design our system to be independent of the availability of annotated datasets as scarcity of such data is expected due to privacy and proprietary concerns. To compensate for annotated datasets, we instead rely on a structured knowledge base in the form of the Unified Medical Language System (UMLS) (Bodenreider, 2004; Lindberg et al., 1993). Taking inspiration from the modular and context-aware frameworks of Phrase-Based Statistical Machine Translation (PBSMT) systems (Koehn et al., 2003), our system, DBF, first identifies hard (low frequency) words such as "SBP", then collects possible simplifications of these words from the UMLS such as {systolic blood pressure, and serotonin-binding protein}, and finally chooses the simplification that best reflects patients' preferred medical terms and best fits the context ("systolic blood pressure" in this case), by relying on a patient language model trained on a suitable monolingual corpus.

Although Neural Machine Translation (NMT) frameworks (Bahdanau et al., 2014; Sutskever et al., 2014) constitute the state-of-the-art, they suffer in the low-resource settings of the clinical (medical) domains, and we accordingly present our system to complement neural methods in domains lacking the appropriate parallel corpus. Although we take medication instructions as a use case, our system

is general enough by construction, to handle any medical text. All code and materials associated with this study are released to the public[1]. This paper makes the following contributions:

- It studies a knowledge-aware text simplification model that does not rely on parallel text.

- The study empirically demonstrates the higher precision simplification output of the proposed model compared to previous methods.

- It makes a parallel corpus of medication instructions available to foster future research.

- It provides a comprehensive and comparative study of NMT models applied to healthcare text simplification, previously impossible due to the lack of a parallel corpus.

- Via a human subjects' study, it shows the positive impact of DBF on patient comprehension.

2 Previous Work

Efforts to improve patient comprehension of health information in the biomedical informatics community can be categorized into: developing standards (Atreja et al., 2005; Wolf et al., 2011), curating dictionaries (Zeng and Tse, 2006), annotating text with additional information (Tupper, 2008; Mohan et al., 2013; Zheng and Yu, 2016; Martin-Hammond and Gilbert, 2016), normalizing terms (Mowery et al., 2016), syntactic simplification (Jonnalagadda and Gonzalez, 2010; Kandula et al., 2010; Peng et al., 2012), and finally, lexical simplification (Chen et al., 2018; Kandula et al., 2010; Qenam et al., 2017). Our work belongs to the final category.

One popular previous attempt (Kandula et al., 2010) of health material text simplification relies on the Consumer Health Vocabulary (CHV) (Zeng and Tse, 2006) for mapping the hard term to its simpler counterpart, disregarding context information. Other word-replacement systems (Chen et al., 2018; Qenam et al., 2017) have relied on MetaMap (Aronson, 2001) to map medical terms to their simpler counterparts by either utilizing CHV as a thesaurus (Qenam et al., 2017), or relying on an in-house equivalent resource (CoDeMed) (Chen et al., 2018). Although, MetaMap performs word sense disambiguation (WSD) by relying on the context, its creators admit its low WSD quality (Aronson and Lang, 2010). Therefore, we rely on a language model instead of MetaMap. Nonetheless, since

---

[1] http://bit.ly/dbf-public-access

MetaMap followed by a CHV (or another dictionary) is a popular method in previous works, we include it as a baseline in our experiments.

Beyond health materials, lexical simplification is highly researched. A thorough survey of this field is present in (Paetzold and Specia, 2017), which divides work in this field into four stages of a pipeline: (1) Complex Word Identification, (2) Substitution Generation, (3) Substitution Selection, and (4) Substitution Ranking. We also perform Complex Word Identification as a first stage by relying on word frequencies, which is a popular method among previous work (Bott et al., 2012; Leroy et al., 2013; Shardlow, 2013; Wróbel, 2016). We also generate substitutions as a second stage by relying on UMLS, similar to how previous work relied on word taxonomies (Carroll et al., 1998; Devlin, 1998). The last two stages are performed in one shot in DBF, where instead of finding which candidate substitutions fit the context and then select the simplest based on a certain metric, we let the language model decide which is the most probable substitution in terms of meaning and simplicity. Our work is the first work to combine these stages in a context-aware method tailored for the healthcare domain. Finally, text simplification has been modeled previously as a machine translation task where parallel corpora are available (Wang et al., 2016; Wubben et al., 2012; Van den Bercken et al., 2019). Accordingly, we compare against these methods in this study to assess their capacity in the low resource setting and their capability to generalize across healthcare domains.

3 Methods

In this section, we describe our method, DBF, along with the established baselines it was quantitatively evaluated against: MetaMap+CHV, Seq2Seq-w-Attention, and Pointer-Generator.

3.1 Dr. Babel Fish

For reproducibility purposes, following is a detailed system description. DBF is designed as a 3-stage pipeline. First, hard (and easy) words are identified based on their frequency of usage. Then, in the second stage, candidate simplifications of a given hard word are collected and each given a replacement probability (p_{rm}). In the final stage, every candidate output simplification is assigned a language model score and a replacement model score. We will refer to this system as an "unsuper-

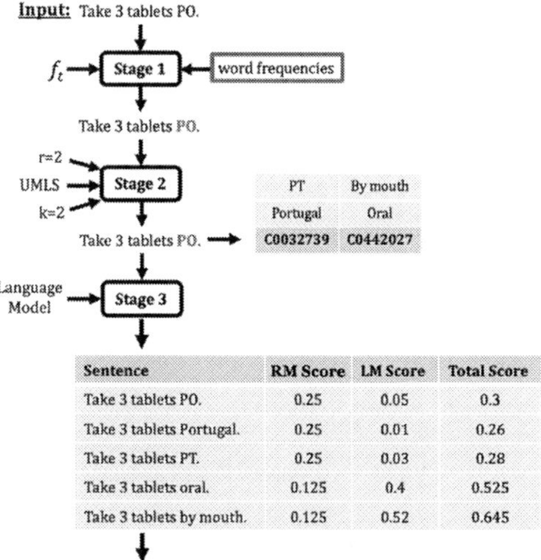

Sentence	RM Score	LM Score	Total Score
Take 3 tablets PO.	0.25	0.05	0.3
Take 3 tablets Portugal.	0.25	0.01	0.26
Take 3 tablets PT.	0.25	0.03	0.28
Take 3 tablets oral.	0.125	0.4	0.525
Take 3 tablets by mouth.	0.125	0.52	0.645

Figure 1: Block diagram of DBF for the sample sentence: "Take 3 tablets PO." RM (replacement model) score represents p_{rm} for the whole system, and LM (language model) score represents p_{lm} for the whole sentence.

vised" system due its independence of annotated datasets, as well as "knowledge-aware" due to its reliance on a knowledge base in the form of UMLS. The highest scoring simplification is then selected as the output of DBF. We describe the 3 stages in the following subsections (see Figure 1).

3.1.1 Stage 1: Identification of Hard Words:

In the first stage, the task is to identify the hard words to be translated from the source sentence and to retain the easy words. Let C_t be a corpus of patient-facing health text. Accordingly, we devise a simple statistical model which checks a word's frequency of usage in C_t. We consider the high usage frequency of a word by patients (or targeted towards patients) to be a strong indicator that it is easy for patients to understand, and vice versa. Thus, if a given word has a frequency lower than a tunable frequency threshold (f_t), DBF labels it as hard.

3.1.2 Stage 2: Candidate Generation:

Next, DBF relies on the UMLS to collect all candidate replacements of each hard word, and estimates the probability of each candidate.

A salient feature of the UMLS is its groupings of words/phrases into clusters, where each cluster represents one concept. In Table 2, we present three example concepts, each headed by its "Pre-

Oral	Twice a day	Milligram
PO	BID	Mg
Orally	Twice daily	Milligramos
By mouth	Two times daily	Milligrams

Table 2: Example UMLS concepts

Query	PO	BID
1st Result	Portugal	BID Protein
2nd Result	Oral	Twice a day
3rd Result	Positive	BID gene

Table 3: Example UMLS queries

ferred Name", followed by three example atoms (the UMLS term for a phrase in a given concept). We note the variability of atoms in a concept in terms of complexity, and language.

A second feature of the UMLS is its ability to return an ordered list of concepts to best match a search query. In Table 3, we see the top three concepts returned for two example queries: "PO", and "BID". The correct concept for "PO" appears only second in the results, as is the case for "BID". This suggests that just relying on the top result of such a context-insensitive static search of the UMLS is insufficient for accurate simplifications.

Leveraging these two features, DBF uses the hard word from the input sentence as a query to the UMLS search function. Then, all atoms of the top k returned concepts are considered as candidate simplifications, with concepts ranked higher assigned higher probabilities.

Formally, let $\{C_1, C_2, ..., C_k\}$ be the top k concepts returned by the UMLS search feature when using the hard word c as a query. Also, let $C_i = \{a_{i1}, a_{i2}, ..., a_{in}\}$ be all the atoms of the ith concept. Then, the probability of atom a_{ij} being the simpler replacement of c is assigned $p_{rm}(a_{ij}|c) \propto \frac{1}{r^i}$, where $r \geq 1$. Thus, an atom of the ith concept is allocated a probability r times that of an atom of the $(i+1)^{th}$ concept. In this setup, r and k are tunable hyperparameters of the system. To allow for possibly keeping a hard word c unaltered on the output side, we also assign probability $p_{rm}(c|c)$ equal to that of an atom in the 1st concept. This helps in cases where a word was wrongly identified as hard, or a simpler alternative does not exist for it. For an easy word e, we assign $p(e|e) = 1$ to force retention of easy words.

3.1.3 Stage 3: Decoding via Language Model:

Finally, we consider all possible combinations of simplifications and choose that with the highest product of replacement probability and language model probability.

Formally, we identify $T(c) = \{t_1, t_2, ..., t_m\}$ which is the set of possible simplifications for a word c. Using this, the set of possible simplifications of the input sentence becomes $H = T(c_1) \times T(c_2) \times T(c_n)$, where $\times$ refers to the Cartesian product of sets.

Now consider a sentence $t \in H$ and let $t = t_1 t_2 ... t_T$ where t_i is the ith word of t. Then, $P(t|c) \propto \prod_{i=1}^{T} p_{rm}(t_i|c_i) * p_{lm}(t_i|t_{i-1:i-5})$, where $p_{lm}(t_i|t_{i-1:i-5})$ is the probability assigned by the 6-gram language model (Brown et al., 1992), for the word t_i occurring after the sequence of words $t_{i-5}t_{i-4}t_{i-3}t_{i-2}t_{i-1}$. The 6-gram language model is trained on the patient-friendly corpus to model the target language. Finally, the sentence t with the highest assigned probability $P(t|c)$ is selected as the output simplification of the system. To further advance this knowledge-aware framework, one can resort to more sophisticated language models such as the recent masked language models (Ghazvininejad et al., 2019).

The significance of the language model is, first, it utilizes the context in which a word like "PO" appears to reward a simplification like "Oral", and penalize a simplification like "Portugal", especially considering that "Portugal" is assigned a higher replacement probability (p_{rm}). Second, it encodes word usage preferences–such as "by mouth" being preferred over "Oral"– even though they both had equal replacement probabilities (p_{rm}).

From an implementation perspective, sentences with high count of hard words would lead to a large H and exponentially slower execution. Hence, we approximate maximizing $P(t|c)$ over H by employing beam search. Traversing H from 1 to n, we maintain only the top 5 sequences as evaluated by $P(t|c)$ up to the respective index.

3.2 MetaMap+CHV:

To compare DBF to the majority of previously used methods for simplifying health materials, we implement the following baseline. Text is first passed through MetaMap, which maps phrases in the text to their respective UMLS concepts. Then, for every phrase identified, we first check if it includes at least one hard word. If it does, and if that UMLS concept is covered by CHV, we replace it by CHV's most preferred term for that concept, otherwise, we replace it with the UMLS preferred term for that concept. With that being said, all phrases identified by MetaMap, which are not contiguous, are ignored to avoid errors in sentence structure when performing the phrase replacement.

3.3 NMT Baselines:

Representing state-of-the-art approaches, our last set of baselines are two supervised NMT architectures (Jhamtani et al., 2017; Sutskever et al., 2014), requiring training data.

One NMT baseline we consider is a Seq2Seq-with-Attention architecture (Sutskever et al., 2014). In this deep learning architecture, a Long Short-Term Memory (LSTM) encoder maps the input sentence to a fixed length vector, and generates contextualized representations of the input words. Then, an LSTM decoder generates the output words sequentially based on the fixed length vector, and the contextualized representations, while the attention mechanism indicates which input words influence each output decision. For this baseline, we utilize Google's open source implementation (Developers, 2017) with default parameters.

Due to the large overlap in the vocabulary of the source and target sentences, particularly the "easy" words, we consider a second NMT baseline called Pointer-Generator capable of copying words as is from the source sentence (Jhamtani et al., 2017). It differs from Seq2Seq-with-Attention in that at every decode step, it estimates a probability g of generating a new word rather than copying a word from the source sentence. If g is low, the model relies more heavily on the estimated attention distribution over the input source words, which increases the chances of copying the word that is most highly weighted by the attention mechanism. To implement the system, we use the author's original open-source implementation (Jhamtani et al., 2017) with default parameters, except for using the Proximal Adagrad (Singer and Duchi, 2009) optimization algorithm to maximize performance.

4 Experiments

This section describes the results of two studies. The first study uses automated evaluation metrics to assess DBF's output in comparison to the baselines considered. The second study evaluates the impact of DBF on laypeople comprehension. We first describe the data used in our experiments and then present the results.

4.1 Data

Parallel Corpus: For the purpose of training supervised NMT methods, and evaluating all systems, we collected 4554 unique and de-identified medication instructions for diabetic patients from the electronic health records of a collaborating healthcare institution. They were of two types: (1) Structured–automatically populated using three drop-down fields: Dose, Route, Frequency (2) Free-text– manually typed. Free-text instructions tend to have more hard words due to their uncontrolled nature.

Then, for every instruction, a physician, with expertise in standard practices for increasing patient comprehension, annotated each instruction with its accurate simplification. Although one other physician was hired for the same task to ensure a high-quality parallel corpus, it was evident by manual inspection, as well as, overall statistics (such as sentence length and word frequencies used in translation), that the former physician's annotations had superior quality. We thus relied solely on the former physician's annotations.

As shown in Table 4, the ratio of structured instructions to free-text ones is around 2:1. On average, structured instructions are slightly lengthier due to the consistency of their length, while free-text instructions vary between the long and short instructions. In terms of novelty—average count of unique new words added in the simplification—simplifications of free-text instructions introduce more novelty due to the complex nature of these instructions. Finally, more free-text instructions were left after simplification. The high level of novelty despite the high number of unchanged instances shows the disparity in instances between long ones that require significant simplification and short ones that require no simplification. We finally check for the average frequency of words (as estimated by the monolingual corpus) in the complicated and simplified side of both types of medication instructions. We notice that, as expected, average frequency increases after simplification. Also, average frequency of words is lower on the free-text side due to their complexity, and the impact of simplification is larger on the free-text side as well. We also include in Appendix A the top 20 most frequent hard words for both types of instructions for a better understanding of the dataset.

	Structured	Free-Text
Instances	3013	1541
Words	11.17	9.94
Novelties	1.44	3.67
Unchanged	200	406
Avg Freq: Compl	68536	59180
Avg Freq: Simpl	69138	65337

Table 4: Parallel Corpus Statistics

Monolingual Corpus: Next, in order to develop a corpus representative of the target language (accessible to patients), we scraped medication-related pages from five medicine-related websites[2] targeted for laypeople. We selected the five websites to be: (1) medication-related, and (2) patient-facing. This corpus C_t ($\approx 11M$ words) was used to: (1) train a language model, and (2) estimate usage frequency of words by DBF's target audience.

Human Subjects' Study: Finally, we designed an online human subjects' study (via Mechanical Turk) that presents medication instructions to participants and tests their comprehension of the instructions, before and after simplification, using multiple-choice questions.

Accordingly, we randomly choose 100 of the free-text medication instructions of varying levels of hardness (1: 29 instructions, 2: 29 instructions, and 3: 42 instructions) as measured by the number of hard (low frequency) words. Then, we simplify every instruction using DBF, and pair both the original and simplified versions of the instruction with the same multiple-choice question.

4.2 Automated Evaluation:

One standard measure for machine translation tasks is BLEU score (Papineni et al., 2002), which measures the overlap in words and phrases between a system's output and a reference output. It is also used frequently in other sequence-to-sequence problems such as text simplification. Nevertheless, BLEU has been shown insufficient for text simplification tasks due to the large overlap between the source and target vocabulary (Xu et al., 2016). Therefore, we instead consider the SARI metric, which showed better correlation than BLEU with human judgement on text simplification tasks (Xu et al., 2016). SARI, similarly to BLEU, measures the overlap of the system's output with a reference output, but also measures the amount of novelty introduced by the system. The novelty component in

the metric rectifies BLEU's shortcoming in measuring the performance of a text simplification system. Moreover, we also use the PINC metric (Chen and Dolan, 2011) to measure, in isolation, the amount of novelty introduced by a system. Readability measures such as the Flesch-Kincaid index (Flesch, 1948), are not suitable for our experiments for at least 2 reasons: (1) they penalize higher word and syllable counts, when most simplifications will increase word and syllable counts such as "PO" to "by mouth", and (2) they are designed for document-level instead of sentence-level assessment.

As for the experimental setup, to avoid evaluating systems on a limited dataset size, we perform 5-fold cross validation to utilize the full dataset for evaluation. For every fold, we take 20% of the training data for tuning.

We present in Table 5 the average performance of all systems on the evaluation portion of the dataset for all five folds. Results of neural baselines were averaged over 3 runs. We also distinguish between the performance on the full dataset and the more critical subset – free-text instructions, and include the results in Table 6. For reference, we also include a baseline system that performs no change. For sample simplifications, the reader is referred to Appendix B.

Method	Supervision Type	PINC	SARI
No Change	N/A	0.00	32.83
MetaMap+CHV	Knowledge-Aware	25.84	45.64
DBF	Knowledge-Aware	19.61	55.33
Pointer-Generator	Direct Supervision	32.25	54.75
Seq2Seq-w-Att	Direct Supervision	50.81	**79.26**

Table 5: Performance of the simplification systems on all medication instructions

Method	Supervision Type	PINC	SARI
No Change	N/A	0.00	39.29
MetaMap+CHV	Knowledge-Aware	26.32	54.35
DBF	Knowledge-Aware	21.52	**56.51**
Pointer-Generator	Direct Supervision	36.34	40.01
Seq2Seq-w-Att	Direct Supervision	78.35	48.27

Table 6: Performance of the simplification systems on the free-text subset of the medication instructions

We first compare the two knowledge-aware systems: MetaMap+CHV, and DBF. First, and confirming our main hypothesis, the context-aware framework of DBF led to higher quality simplifications gaining an absolute 9.7% improvement in SARI scores over MetaMap+CHV, and a 22.5%

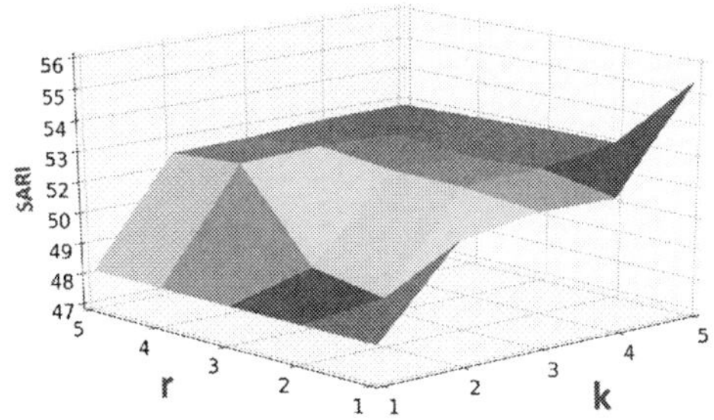

Figure 2: Effect of the hyperparameters k and r on the performance of DBF

gain compared to the No-Change case. Even though MetaMap has the added flexibility to operate on a phrase level, we attribute its comparatively lower quality to its poor WSD. Second, and by comparing PINC scores, we notice that DBF is more conservative in its changes, making it less likely to mistakenly alter key information, arguably a desired behavior in a critical domain such as healthcare. This is mainly due to it considering the identity replacement as a possible simplification, and letting the context decide whether to attempt simplification or not. These observations are also consistent on the free-text subset of the evaluation data, though we note that the gap shrinks between the two systems. We hypothesize that this is due to MetaMap+CHV committing consistent errors over one or more highly repeated terms in the structured subset of the medication instructions.

Next, we observe that, including the supervised deep learning methods, Seq2Seq-w-Attention performs significantly better than all systems. The high performance of the Seq2Seq-w-Attention is an expected result, due to the advantage of direct supervision in general, but also because direct supervision would allow it to memorize the annotator's style as well. The poor performance of the Pointer-Generator was unexpected considering its mechanism to pass easy words. Upon further inspection, we noticed two factors that degraded performance. First, the copy mechanism led to meaningless repetition of words as previously noted in the literature (See et al., 2017). Second, the copy mechanism led to copying hard words as is.

Finally, we focus our attention on how performance levels are affected when considering free-text instructions only, which are more representative of complicated health material. We notice that all the systems show more activity (higher PINC scores) in their simplifications, as these sys-tems encounter more hard words in the original instructions. This provides further evidence that the free-text instructions constitute a critical component of the evaluation. Second, we notice that the performance of the supervised systems suffers significantly on the free-text instructions (compared to that on All Instructions), while that of the knowledge-aware (utilizing background knowledge such as UMLS and CHV) systems remain comparable, to the extent that DBF becomes the best performing approach on free-text instructions. This reflects the robustness of the knowledge-aware systems in a low-resource setting. In a setting where a sufficient parallel corpus is available, neural machine translation systems are recommended, most notably Transformer-based (Vaswani et al., 2017) for future endeavours. But in the absence of a sizable in-domain corpus, DBF achieves better performance.

Furthermore, it is notable that DBF was the best performing system despite it being the only one limited to lexical simplification. A concrete future direction would be to extend DBF's capabilities to perform phrase-level simplification.

4.3 Simplification Effects on Patient Comprehension:

We also investigated whether DBF's simplification efficacy helped improve laypeople comprehension, by measuring their ability to answer multiple choice questions (percent correct) on medication instructions before and after simplification (see Figure 3). Participants, on Amazon Mechanical Turk, were 160 adults diverse in age, cultural and academic background, and gender. 100 instructions were randomly selected from the free-text subsample of our original set of medication instructions (see Materials Section), along with their DBF simplifications and their respective multiple choice questions. Each participant read 50 instructions and answered the corresponding questions. A counterbalancing scheme ensured that each participant read 25 original instructions (as written by the physician) and 25 instructions simplified by DBF. No participant encountered both the original and the simplified version of the same instruction. Also, hardness levels of medication instructions were balanced for each participant.

The key result of this experiment was that the participants understood the simplified instructions 24.4% better than the original instructions

Medication Instruction: 10 mg PO 6pm daily.

Question: How should you take this medicine?

a) By needle
b) Into the butt hole
c) By mouth
d) On the skin
e) It was not indicated in the medication instruction

Medication Instruction: 10 mg orally 6pm daily.

Question: How should you take this medicine?

a) By needle
b) Into the butt hole
c) By mouth
d) On the skin
e) It was not indicated in the medication instruction

Figure 3: Example questions from the online human subjects study

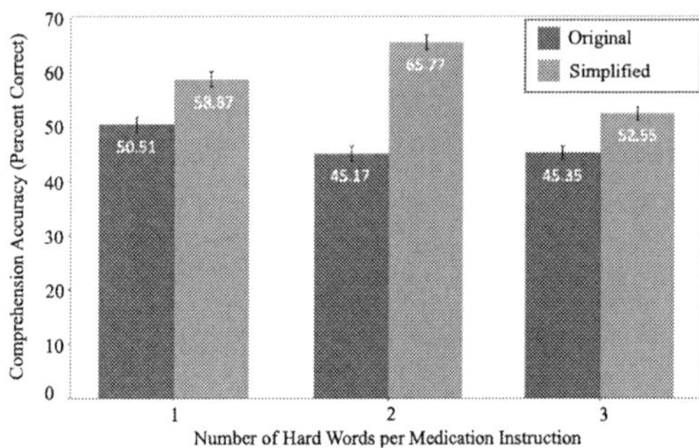

Figure 4: Impact of DBF on the different hardness levels of medication instructions

$(F(1, 7973) = 112.3, p < .0001, 58.23\%$ vs $46.80\%)$. Hardness level also influenced comprehension $(F(2, 7973) = 14.6, p < .0001$; see Figure 4). The simplification benefit was largest when there were two difficult words (45.63% relative), rather than one (16.55% relative) or three difficult words (15.87% relative). It is possible that having two rather than one difficult word gave more potential for DBF's simplification to increase comprehension. However, when simplification involved three words, the propagation of error led to a decrease in the quality of the simplification, and this may have negatively impacted comprehension.

5 Discussion

To better understand the functioning of the knowledge-aware systems, we study the effect of f_t on their first stage of identifying hard words. Upon tuning the systems on the validation dataset, f_t was set to 672 for both systems coincidentally. Based on Figure 5, we deduce that around 18% of words in the original instructions were attempted for translation, reflecting a high recall of hard words.

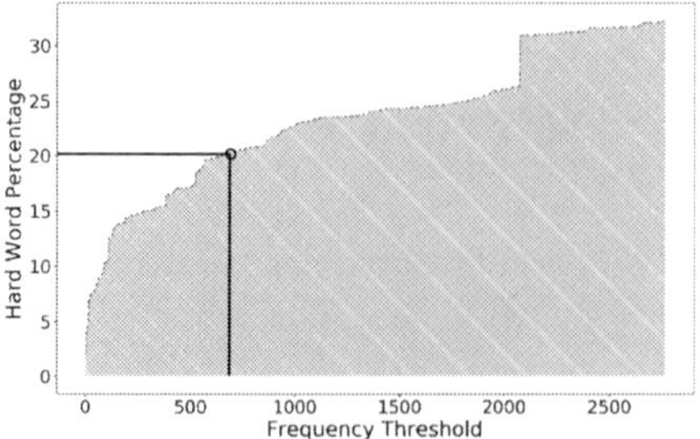

Figure 5: Percentage of words considered hard as we vary the frequency threshold

Along the same lines, we check the effect of f_t on DBF and MetaMap+CHV in terms of the two evaluation scores (see Figure 6). In terms of PINC scores, we observe an expected pattern of increase as we increase f_t for both systems. As f_t increases, both systems attempt to modify more of the original sentence (including easy words) leading to a lower overlap with reference sentences. MetaMap+CHV introduces more novelty as we increase f_t since DBF can retain easy words even if they were identified as hard, unlike MetaMap+CHV. As for SARI scores, we observe the significance of tuning the first stage, where too low of an f_t results in reduced performance due to lack of attempted translations (low PINC scores), and too high of an f_t results in reduced performance due to translating easy words. Moreover, we observe consistent enhances in performance for DBF over MetaMap+CHV for all f_t considered.

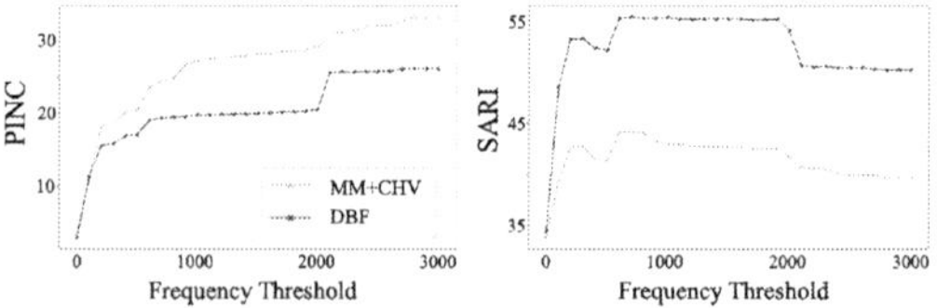

Figure 6: Effect of the frequency threshold on the performance of DBF, and MetaMap+CHV

Moving our attention to the effect of k and r on the performance of DBF, we show in Figure 2, DBF's SARI score when varying k and r from 1 to 5, and fixing f_t to 672. Our first observation is a positive trend as we increase k, particularly for $r = 1$. This shows the aptitude of the language model at selecting the best translation even when faced with a plethora of options given equal translation probabilities. As for r, we notice reduced performances for any r value different from 1. We thus conclude that the model we use for estimating translation probabilities is not benefiting translation quality. Moreover, the ranking of the concepts returned by the UMLS search function has insignificant value, when an appropriate language model is present.

6   Conclusion

Despite significant efforts to keep patients informed of their health condition, the gap in health literacy remains an issue, calling for a necessary text simplification system to bridge the gap. In this work, we suggest a context-aware framework to ensure high accuracy in such a critical domain, while also showing its positive impact on comprehension through a human subjects' study. We also conclude that, while supervised NMT methods are well-suited for the task, several healthcare subdomains lack suitable parallel corpora, which limits the performance of these supervised methods. To overcome this, we offer a knowledge-aware text simplification system to robustly operate in a low-resource setting.

Looking forward, we see great potential in adapting deep learning architectures to utilize the rich and highly curated content of UMLS, and exploring better methods or implementations to copy easy words to the target side. These two adaptations would make the high performing supervised methods of NMT less dependent on direct supervision and more generally applicable to the multiple healthcare domains. This would also address the limitation of DBF to word-level simplifications.

Acknowledgment

This work was partly supported by the IBM-ILLINOIS Center for Cognitive Computing Systems Research (C3SR) — a research collaboration as part of the IBM AI Horizons Network. This work was also partly supported by the Jump Applied Research for Community Health through Engineering and Simulation (ARCHES) program, UIUC/OSF Hospital, Peoria IL.

References

Alan R Aronson. 2001. Effective mapping of biomedical text to the umls metathesaurus: the metamap program. In *Proceedings of the AMIA Symposium*, page 17. American Medical Informatics Association.

Alan R Aronson and François-Michel Lang. 2010. An overview of metamap: historical perspective and recent advances. *Journal of the American Medical Informatics Association*, 17(3):229–236.

Ashish Atreja, Naresh Bellam, and Susan R Levy. 2005. Strategies to enhance patient adherence: making it simple. *Medscape General Medicine*, 7(1):4.

Dzmitry Bahdanau, Kyunghyun Cho, and Yoshua Bengio. 2014. Neural machine translation by jointly learning to align and translate. *arXiv preprint arXiv:1409.0473*.

Laurens Van den Bercken, Robert-Jan Sips, and Christoph Lofi. 2019. Evaluating neural text simplification in the medical domain. In *The World Wide Web Conference*, pages 3286–3292.

Olivier Bodenreider. 2004. The unified medical language system (umls): integrating biomedical terminology. *Nucleic acids research*, 32(suppl_1):D267–D270.

Stefan Bott, Luz Rello, Biljana Drndarević, and Horacio Saggion. 2012. Can spanish be simpler? lexis: Lexical simplification for spanish. In *Proceedings of COLING 2012*, pages 357–374.

Peter F Brown, Peter V Desouza, Robert L Mercer, Vincent J Della Pietra, and Jenifer C Lai. 1992. Class-based n-gram models of natural language. *Computational linguistics*, 18(4):467–479.

John Carroll, Guido Minnen, Yvonne Canning, Siobhan Devlin, and John Tait. 1998. Practical simplification of english newspaper text to assist aphasic readers. In *Proceedings of the AAAI-98 Workshop on Integrating Artificial Intelligence and Assistive Technology*, pages 7–10.

David L Chen and William B Dolan. 2011. Collecting highly parallel data for paraphrase evaluation. In *Proceedings of the 49th Annual Meeting of the Association for Computational Linguistics: Human Language Technologies-Volume 1*, pages 190–200. Association for Computational Linguistics.

Jinying Chen, Emily Druhl, Balaji Polepalli Ramesh, Thomas K Houston, Cynthia A Brandt, Donna M Zulman, Varsha G Vimalananda, Samir Malkani, and Hong Yu. 2018. A natural language processing system that links medical terms in electronic health record notes to lay definitions: System development using physician reviews. *Journal of medical Internet research*, 20(1):e26.

Karen Davis, Stephen C Schoenbaum, and Anne-Marie Audet. 2005. A 2020 vision of patient-centered primary care. *Journal of general internal medicine*, 20(10):953–957.

Don Detmer, Meryl Bloomrosen, Brian Raymond, and Paul Tang. 2008. Integrated personal health records: transformative tools for consumer-centric care. *BMC medical informatics and decision making*, 8(1):45.

TensorFlow Developers. 2017. Tensorflow neural machine translation tutorial.

Siobhan Devlin. 1998. The use of a psycholinguistic database in the simplification of text for aphasic readers. *Linguistic databases*.

Rudolph Flesch. 1948. A new readability yardstick. *Journal of applied psychology*, 32(3):221.

Marjan Ghazvininejad, Omer Levy, Yinhan Liu, and Luke Zettlemoyer. 2019. Mask-predict: Parallel decoding of conditional masked language models. *arXiv preprint arXiv:1904.09324*.

Jennifer Fong Ha and Nancy Longnecker. 2010. Doctor-patient communication: a review. *Ochsner Journal*, 10(1):38–43.

Taya Irizarry, Annette DeVito Dabbs, and Christine R Curran. 2015. Patient portals and patient engagement: a state of the science review. *Journal of medical Internet research*, 17(6):e148.

Harsh Jhamtani, Varun Gangal, Eduard Hovy, and Eric Nyberg. 2017. Shakespearizing modern language using copy-enriched sequence-to-sequence models. *arXiv preprint arXiv:1707.01161*.

Siddhartha Jonnalagadda and Graciela Gonzalez. 2010. Biosimplify: an open source sentence simplification engine to improve recall in automatic biomedical information extraction. In *AMIA Annual Symposium Proceedings*, volume 2010, page 351. American Medical Informatics Association.

Sasikiran Kandula, Dorothy Curtis, and Qing Zeng-Treitler. 2010. A semantic and syntactic text simplification tool for health content. In *AMIA annual symposium proceedings*, volume 2010, page 366. American Medical Informatics Association.

Roy PC Kessels. 2003. Patients' memory for medical information. *Journal of the Royal Society of Medicine*, 96(5):219–222.

David A Kindig, Allison M Panzer, Lynn Nielsen-Bohlman, et al. 2004. *Health literacy: a prescription to end confusion*. National Academies Press.

Philipp Koehn, Franz Josef Och, and Daniel Marcu. 2003. Statistical phrase-based translation. In *Proceedings of the 2003 Conference of the North American Chapter of the Association for Computational Linguistics on Human Language Technology-Volume 1*, pages 48–54. Association for Computational Linguistics.

Mark Kutner, Elizabeth Greenburg, Ying Jin, and Christine Paulsen. 2006. The health literacy of america's adults: Results from the 2003 national assessment of adult literacy. nces 2006-483. *National Center for Education Statistics*.

Gondy Leroy, James E Endicott, David Kauchak, Obay Mouradi, and Melissa Just. 2013. User evaluation of the effects of a text simplification algorithm using term familiarity on perception, understanding, learning, and information retention. *Journal of medical Internet research*, 15(7):e144.

Donald AB Lindberg, Betsy L Humphreys, and Alexa T McCray. 1993. The unified medical language system. *Yearbook of Medical Informatics*, 2(01):41–51.

Aqueasha Martin-Hammond and Juan E Gilbert. 2016. Examining the effect of automated health explanations on older adults' attitudes toward medication information. In *Proceedings of the 10th EAI International Conference on Pervasive Computing Technologies for Healthcare*, pages 186–193.

Nicholas Mcinnes and Bo JA Haglund. 2011. Readability of online health information: implications for health literacy. *Informatics for health and social care*, 36(4):173–189.

Arun V Mohan, M Brian Riley, Dane R Boyington, and Sunil Kripalani. 2013. Illustrated medication instructions as a strategy to improve medication management among latinos: a qualitative analysis. *Journal of health psychology*, 18(2):187–197.

Danielle L Mowery, Brett R South, Lee Christensen, Jianwei Leng, Laura-Maria Peltonen, Sanna Salanterä, Hanna Suominen, David Martinez, Sumithra Velupillai, Noémie Elhadad, et al. 2016. Normalizing acronyms and abbreviations to aid patient understanding of clinical texts: Share/clef ehealth challenge 2013, task 2. *Journal of biomedical semantics*, 7(1):43.

Gustavo H Paetzold and Lucia Specia. 2017. A survey on lexical simplification. *Journal of Artificial Intelligence Research*, 60:549–593.

Kishore Papineni, Salim Roukos, Todd Ward, and Wei-Jing Zhu. 2002. Bleu: a method for automatic evaluation of machine translation. In *Proceedings of the 40th annual meeting on association for computational linguistics*, pages 311–318. Association for Computational Linguistics.

Yifan Peng, Catalina O Tudor, Manabu Torii, Cathy H Wu, and K Vijay-Shanker. 2012. isimp: A sentence simplification system for biomedicail text. In *2012 IEEE International Conference on Bioinformatics and Biomedicine*, pages 1–6. IEEE.

Basel Qenam, Tae Youn Kim, Mark J Carroll, and Michael Hogarth. 2017. Text simplification using consumer health vocabulary to generate patient-centered radiology reporting: translation and evaluation. *Journal of medical Internet research*, 19(12):e417.

AK Rotegard, Laura Slaughter, and Cornelia M Ruland. 2006. Mapping nurses' natural language to oncology patients' symptom expressions. *Studies in health technology and informatics*, 122:987.

Abigail See, Peter J Liu, and Christopher D Manning. 2017. Get to the point: Summarization with pointer-generator networks. *arXiv preprint arXiv:1704.04368.*

Matthew Shardlow. 2013. The cw corpus: A new resource for evaluating the identification of complex words. In *Proceedings of the Second Workshop on Predicting and Improving Text Readability for Target Reader Populations*, pages 69–77.

Yoram Singer and John C Duchi. 2009. Efficient learning using forward-backward splitting. In *Advances in Neural Information Processing Systems*, pages 495–503.

Ilya Sutskever, Oriol Vinyals, and Quoc V Le. 2014. Sequence to sequence learning with neural networks. In *Advances in neural information processing systems*, pages 3104–3112.

J RC Tupper. 2008. Plain language thesaurus for health communications.

Ashish Vaswani, Noam Shazeer, Niki Parmar, Jakob Uszkoreit, Llion Jones, Aidan N Gomez, Łukasz Kaiser, and Illia Polosukhin. 2017. Attention is all you need. In *Advances in neural information processing systems*, pages 5998–6008.

Tong Wang, Ping Chen, John Rochford, and Jipeng Qiang. 2016. Text simplification using neural machine translation. In *Thirtieth AAAI Conference on Artificial Intelligence.*

Michael S Wolf, Laura M Curtis, Katherine Waite, Stacy Cooper Bailey, Laurie A Hedlund, Terry C Davis, William H Shrank, Ruth M Parker, and Alastair JJ Wood. 2011. Helping patients simplify and safely use complex prescription regimens. *Archives of internal medicine*, 171(4):300–305.

Krzysztof Wróbel. 2016. Plujagh at semeval-2016 task 11: Simple system for complex word identification. In *Proceedings of the 10th International Workshop on Semantic Evaluation (SemEval-2016)*, pages 953–957.

Sander Wubben, Antal Van Den Bosch, and Emiel Krahmer. 2012. Sentence simplification by monolingual machine translation. In *Proceedings of the 50th Annual Meeting of the Association for Computational Linguistics: Long Papers-Volume 1*, pages 1015–1024. Association for Computational Linguistics.

Wei Xu, Courtney Napoles, Ellie Pavlick, Quanze Chen, and Chris Callison-Burch. 2016. Optimizing statistical machine translation for text simplification. *Transactions of the Association for Computational Linguistics*, 4:401–415.

Qing T Zeng and Tony Tse. 2006. Exploring and developing consumer health vocabularies. *Journal of the American Medical Informatics Association*, 13(1):24–29.

Structured	Count	Free-text	Count
route	711	indications	332
tab	439	units	160
ml	319	tab	129
tabs	286	tabs	115
subcutaneous	228	x	108
nightly	191	po	97
units	164	evening	38
cap	149	friday	36
g	115	monday	35
nebulization	78	am	31
evening	74	wednesday	27
mcg	68	ml	26
inhalation	68	scale	25
caps	59	q	24
admin	46	sunday	23
spasms	40	saturday	23
puffs	38	sliding	22
wheezing	35	pt	22
transdermal	28	bedtime	21
breakfast	27	tuesday	21

Table 7: Top 20 Most frequent words identified by our system as hard for structured and free-text medication instructions

Jiaping Zheng and Hong Yu. 2016. Methods for linking ehr notes to education materials. *Information Retrieval Journal*, 19(1-2):174–188.

A Hard Words

To better understand the dataset and the transformations DBF performs, we include in Table 7 the top 20 most frequent hard words in both structured and free-text medication instructions. A word is identified as hard if its frequency in the monolingual corpus C_t is less than the tuned hyperparameter $f_t = 672$.

On the structured side, we notice a concentration of words describing the route: "subcutaneous", "nebulization", "inhalation", "transdermal". We also find a concentration of units: "ml", "g", "mcg". On the free-text side, we observe an abundance of weekdays, as well as non-standard abbreviations such as: "po" for "by mouth", "pt" for "patient", and "q" for "every".

B Example Simplifications

Here, we analyze several example simplifications from the various systems in Table 8. The first example shows the incapability of Seq2Seq-w-Att to recover from a wrongly generated

Source:	Total 90 mg QAM.
Gold:	Total 90 milligrams every morning.
DBF:	Total 90 mg every morning.
MetaMap+CHV:	Total 90 mg every morning.
Seq2Seq-w-Att:	Wheeled systolic blood sugar test result is between 301 and 180,
Pointer-Generator:	Total 90 mg QAM.
Source:	Every 4-6 hours PRN thoracic back pain.
Gold:	Every 4 up to 6 hours as needed for chest back pain.
DBF:	Every 4-6 hours as needed thoracic back pain.
MetaMap+CHV:	Every 4-6 hours PRN thoracic back pain.
Seq2Seq-w-Att:	Every 6 hours as needed for back pain.
Pointer-Generator:	Every 4-6 hours PRN back pain.
Source:	Take 15 g by mouth 2 times daily as needed.
Gold:	Take 15 grams by mouth 2 times daily as needed.
DBF:	Take 15 grams by mouth 2 times daily as needed.
MetaMap+CHV:	Take 15 gram per deciliter by mouth 2 times daily as needed.
Seq2Seq-w-Att:	Take 15 grams by mouth 2 times daily as needed.
Pointer-Generator:	Take 15 grams by mouth 2 times daily as needed.
Source:	For better hearing with the ear, avoid cleaning your cerumen.
Gold:	For better hearing with the ear, avoid cleaning your earwax.
DBF:	For better hearing with the ear, avoid cleaning your wax.
MetaMap+CHV:	For better hearing with the ear, avoid cleaning your earwax.
Seq2Seq-w-Att:	Provide syringes dressings with the month, and Sunday more Lantus.
Pointer-Generator:	For UNK UNK with the UNK UNK

Table 8: Sample output simplifications from the different systems considered

first word (Wheeled). Moreover, we notice Pointer-Generator's tendency to even pass hard words. In the second example, we notice how MetaMap+CHV retains "PRN" despite being a hard word, due to MetaMap not mapping it to any UMLS concept. Additionally, we see Seq2Seq-w-Attention's mishandling of numbers since it does not have a mechanism for passing easy words. We also notice how Pointer-Generator wrongly eliminates words (thoracic) essential to the meaning of the sentence. On the other hand, the next example shows the shortcomings of MetaMap+CHV's disambiguation algorithms, while DBF was able to accurately map "g" to "grams". Whereas both deep learning methods get the full mark on this example since it is a structured medication instruction.

The last point we would like to address is the last example in Table 8. This example, contrary to the previous ones, was not taken from the medication instruction dataset, but rather created by us to portray a complicated sentence from another medical domain, in this case: online health tips. As can be seen from the systems' outputs, the robustness of knowledge-aware systems is evident in comparison to the supervised deep learning methods, which are completely off the mark.

Identifying Personal Experience Tweets of Medication Effects Using Pre-trained RoBERTa Language Model and Its Updating

Minghao Zhu[1,2], Youzhe Song[1,2], Ge Jin[2], Keyuan Jiang[2]
[1]Donghua University, Shanghai, China
[2]Purdue University Northwest, Hammond, Indiana, U.S.A.
minghao.zhu0@gmail.com, isidoresongsisidoresongs@gmail.com, jin9@pnw.edu, kjiang@pnw.edu

Abstract

Post-market surveillance, the practice of monitoring the safe use of pharmaceutical drugs is an important part of pharmacovigilance. Being able to collect personal experience related to pharmaceutical product use could help us gain insight into how the human body reacts to different medications. Twitter, a popular social media service, is being considered as an important alternative data source for collecting personal experience information with medications. Identifying personal experience tweets is a challenging classification task in natural language processing. In this study, we utilized three methods based on Facebook's Robustly Optimized BERT Pretraining Approach (RoBERTa) to predict personal experience tweets related to medication use: the first one combines the pre-trained RoBERTa model with a classifier, the second combines the updated pre-trained RoBERTa model using a corpus of unlabeled tweets with a classifier, and the third combines the RoBERTa model that was trained with our unlabeled tweets from scratch with the classifier too. Our results show that all of these approaches outperform the published methods (Word Embedding + LSTM) in classification performance ($p < 0.05$), and updating the pre-trained language model with tweets related to medications could even improve the performance further.

1 Introduction

Personal experience is an important piece of information for health-related surveillance activities. Understanding one's health experience can help gain insight into the status of one's health, changes of one's health condition after the intervention, or the effects related to any medications one took.

Investigating effects related to the use of pharmaceutical products is an important activity of post-market surveillance. First-hand information related to patients' medication use most directly reflects the effects of the medication, beneficially or adversely. In that case, it is necessary to find valuable data sources and construct efficient methods for processing and analyzing this data.

The widespread availability of social media has made it possible for people to share their personal experiences freely online. Twitter is one of the most prevalent social media services, and studies have shown that the data from social media such as Twitter has been applied to many health-related applications. Examples are as follows: drug adverse events (Bian et al. 2012), public health (Paul et al. 2011; Parker et al. 2013), mental health (Coppersmith et al. 2014; Reece et al. 2017), dental pain (Heaivilin et al. 2011), influenza (Lee et al. 2013; Paul et al. 2015; Gesualdo et al. 2013; Aramaki et al. 2011; Byrd et al. 2016; Kagashe et al. 2017), breast cancer (Thackeray et al. 2013), and epidemic outbreak and spread detection (Ji et al. 2012).

Personal experience is about a person's encounters or observations related to his or her life. Personal experience information related to the use of medication is of unique value for post-market surveillance because it is the first-hand information that reflects the health condition changes due to medication usage. Personal Experience Tweets (PETs) related to medication use are a kind of Twitter post expressing one's personal experience and information after the administration of medication. The types of experiences could be undesirable feelings caused by medications' side-effects, or beneficial effects that help improve a

Proceedings of the 11th International Workshop on Health Text Mining and Information Analysis, pages 127–137
November 20, 2020. ©2020 Association for Computational Linguistics
https://doi.org/10.18653/v1/P17

medication user's health condition. The collection and understanding of these experiences' information can help promote the safe use of medications and advance our healthcare practices. Here are some examples of PETs related to medication use (the underscored text is for medication effects and the boldfaced for the medication):

*"Slow release **morphine** <u>almost killed me</u>."*

*"my mother <u>developed bleeding ulcers</u> from **naproxen** and now they switched her to celebrex isnt that just as bad?"*

*"Ill check it out - I have a friend on **Abilify** and <u>hes had some personality changes</u>, IE <u>agitation</u>, <u>hitting stuff</u>, ect."*

These tweets show that the effects are associated with a person's experience. In contrast, we define a tweet not describing a personal experience as a non-PET. The following are some examples:

*"wish i had some **xanax** to put me to <u>sleep</u>"*

*"**ativan** please help me <u>get some sleep</u> tonight"*

*"i just took a dose of **percocet** with some strippers"*

The above non-PETs, albeit mentioning medications or containing effect expressions, do not reflect the personal experience.

Extracting PETs from various kinds of Twitter posts is challenging because the Twitter data is of abundant noises, and most of the tweets may be irrelevant to personal experience about health conditions. In addition, users usually post tweets with informal and casual styles, without following the rules of grammar and/or spelling. Finally, Twitter users are creative in coining short text to include the needed information within the space limit. These unique characteristics make it more challenging to identify PETs accurately.

2   Related Works

Distinguishing PETs and non-PETs can be treated as a binary classification task. In the conventional machine learning field, algorithms require a set of manually engineered features extracted from the raw text and/or metadata (Jiang et al., 2016; Wijeratne et al., 2017), usually known as feature engineering, and features chosen can significantly impact the classifier's performance. However, extracting/engineering valuable yet optimal features from tweets is difficult due to the limitation of human knowledge and understanding even for the domain experts. Besides, feature engineering extracts features that are typically based on the analysis of statistics regarding information gain usually with little or no direct consideration of the semantics. In other words, conventional machine learning with feature engineering methods may not be optimal for this task.

Efforts of performance improvement have been made in previous research endeavors in the task of predicting personal experience tweets related to medication effects. In one of the earliest efforts, personal pronouns were considered as an important feature (Jiang and Zheng, 2013). Later, Alvaro and colleagues engineered a set of features (Alvaro et al., 2015), and their features include Twitter-specific features, n-grams, punctuation elements, and topics, but the group decided to discard the topic feature due to the significant efforts required and its minimum merit of improving classification performance. A set of 22 engineered features based upon both textual content and metadata of tweets was proposed in constructing a corpus of personal experience tweets (Jiang et al., 2016). Subsequently, Calix and colleagues introduced the concept of deep gramulator to include a textual feature that contains expressions in one class but not in the opposite class, to improve the discriminatory ability of the classification (Calix et al., 2017). Advancement in neural embedding, which demonstrated state-of-art results in many classification tasks on textual data, motivated the development of a new approach of combining word embedding (word2vec) and a recurrent neural network which demonstrated a significant improvement of classification performance ($p <$ 0.05) (Jiang et al., 2018).

Thanks to the development of word embedding techniques and the long short term memory (LSTM) neural network, Jiang et al. (2019) assessed a set of different word embedding techniques: GloVe (Pennington et al. 2014), fastText (Bojanowski et al. 2016) and word2vec (Mikolov et al. 2013) to build vector space models (VSM) to represent the semantics of tweets by learning from a corpus of 22 million unlabeled tweets. The vector representations of tweets were fed into an LSTM neural network for classification. All of these methods achieved better performance in classification measures than the previous methods with 22 human-engineered features using conventional classification algorithms (Jiang et al. 2016).

Unlike the word embedding + LSTM method, which need to learn the VSM first and then train the LSTM network from scratch for classification, Google introduced a fine-tuning based approach by proposing the Bidirectional Encoder Representations from Transformers (BERT) model (Devlin et al. 2018), which achieved record-breaking results in 18 downstream NLP tasks. Besides, Google's new method relies on contextual information rather than term co-occurrences. After that, Facebook made some optimization based on BERT and released a Robustly Optimized BERT Pretraining Approach (RoBERTa) model (Liu et al. 2019) which generated even better performance than BERT in downstream tasks. One important and useful aspect of both approaches is that the pre-trained models can be updated with new data, without the need to generate a new model from scratch with the added data, which generally requires a significant amount of computation resources.

In this study, we set the performance of the word embedding + LSTM neural network method as the baseline and investigated the performance improvements of PETs prediction with the pre-trained RoBERTa language model. We also studied a procedure of updating the pre-trained RoBERTa language model and training the RoBERTa from scratch with the medication-related tweets and analyzing the impact on the performance change.

3 Method

In this work, we introduced three ways to identify personal experience tweets about medication effects by using RoBERTa language model: (1) Pretrained RoBERTa - adding a classifier to the standard pre-trained RoBERTa model and fine-tuning the model for classification; (2) Updated RoBERTa - updating the pre-trained RoBERTa language model with our dataset first, then adding a classifier to RoBERTa and fine-tuning the model for classification; and (3) Twitter RoBERTa - training the RoBERTa language model with our corpus of unannotated tweets from scratch, then adding a classifier for classification. Finally, 10-fold cross-validation was performed to gather the performance data, and statistical analysis was performed to determine if the differences in performance among different methods were due to the chance.

The pipelines of data processing and analysis is illustrated in Figure 1. Our process started with

gathering Twitter data and performing text encoding after preprocessing. Afterwards, the encoded texts were used with the RoBERTa model and the classifier for our methods. The left pipeline is for the Pretrained RoBERTa approach, the middle one for Updated RoBERTa, and the right one for Twitter RoBERTa.

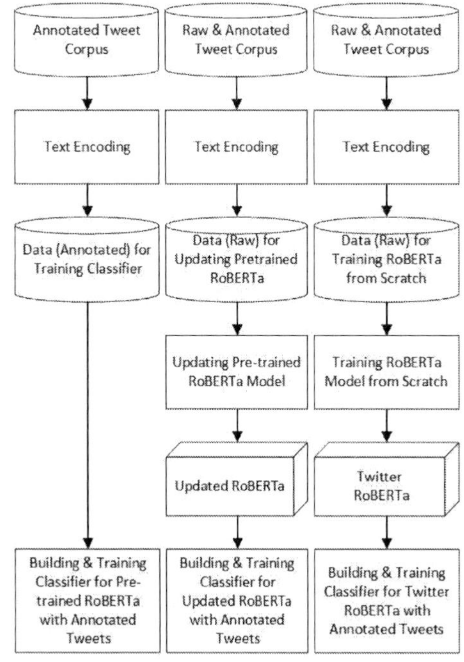

Figure 1. The pipelines of data processing.

3.1 Text Encoding

Byte-Pair Encoding (BPE) (Sennrich et al. 2015) and Attention Mask were applied to encode raw text. BPE is a sub-word level encoding method that uses bytes as the base sub-word units. In the process of tokenization, tokens like acronyms, abbreviations or spelling mistakes which are not in the vocabulary are split into known sub-word tokens, Compared to the word-level encoding method, it is flexible enough for tokenized words with special forms and adaptable for most of English documents, and also it could efficiently avoid most of the unknown tokens in the input text. A sub-word vocabulary with 50K unique tokens was built before pre-training, which was tested with our dataset to ensure that our data could be completely covered by this vocabulary and tweet text was tokenized properly without leaving any unknown tokens. In that case, we reused this sub-word vocabulary to encode our data and each of the tweets was converted into a sequence of indices of tokens in the vocabulary.

After encoding, each tweet started with a special <s> token and ended with </s>. To achieve the fixed length of a sequence, we set the max token length to 64, and a special <pad> token was introduced to pad sequences to the max length. We ensured that this value of max token length could fit almost all of the tweets: only 0.003% of them were longer than 64 tokens. Also, an Attention Mask was applied to all of the input data to avoid performing attention on padding tokens. For each sentence, 0 is for padding tokens that should be masked, and 1 is for others that are not masked. Figure 2 shows an example of text encoding.

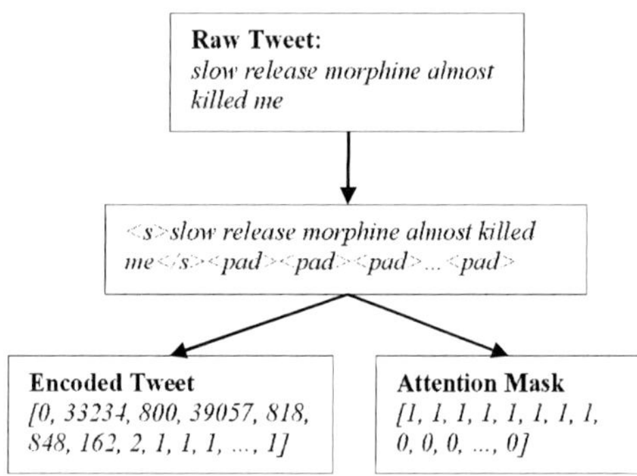

Figure 2. Example of text encoding

3.2 Pre-training

Pre-training the language model in a large corpus could help the model learn a series of general common properties of the language, and it is expected to be used in some of the downstream target tasks with a small dataset where it could perform better. The pre-training model we used is based upon the model of RoBERTa, whose structure is based on Google's BERT model, with 12 layers, 768 hidden neurons, 12 self-attention heads and a total of 110M parameters. The RoBERTa model was released by Facebook AI (Liu et al. 2019), pre-trained with masked language model (MLM) task: 15% of tokens were randomly and dynamically selected for replacement; 80% of them were replaced by a special token <mask>; 10% were kept unchanged; the rest of 10% of the tokens were replaced by a random token in vocabulary. The pre-training procedure was performed on a total of over 160GB uncompressed texts for 500K steps with an 8K batch size.

3.3 Language Model (LM) Updating

Although the pre-trained model extracts the general features of linguistic expression in a large corpus, the dataset of our task could be in a different distribution. To make the pre-trained model adapt to our task, we updated the pre-trained RoBERTa model with our corpus of 10M unlabeled tweets before training the classifier. In this updating procedure, we implemented the same masking strategy as that of the masked LM task in the pre-training procedure, described previously, with a set of newly designated hyperparameters *(training steps: 53K/106K/160K batch size: 64, optimizer: Adam, learning rate 2×10^{-5})*

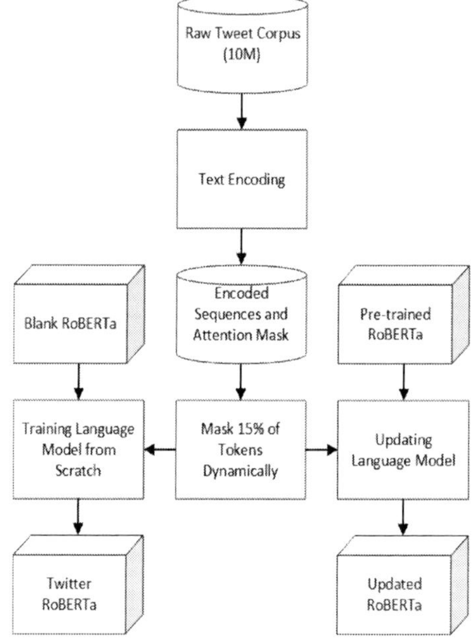

Figure 3. Setup of Updated RoBERTa and Twitter RoBERTa

3.4 Training RoBERTa from Scratch

Another way to let the model learn the property and distribution of a new language environment is to train a new model from scratch with the new dataset. As for our task, it is also a selectable approach. To determine whether training the RoBERTa model with our corpus of tweets could perform better than Facebook's pre-trained one and to use the updating approach in predicting personal experience tweets, a new Twitter RoBERTa model was constructed with the same corpus of tweets as the updating procedure use. Due to the hardware difference between Facebook's and ours, a set of different hyperparameters were used to train it from scratch. *(training steps: 53K/106K/160K, optimizer: Adam, learning rate: 5×10^{-5}, batch size: 64)* Figure 3 illustrates the overview of the procedure of LM updating and training the Twitter RoBERTa from scratch.

3.5 Classifier Fine-tuning

A classifier with a simple feedforward neural network was constructed by following RoBERTa's original design, which is adapted for RoBERTa's base concepts and structure. This is also officially recommended to use for the most of downstream classification tasks by Facebook AI. The classifier is made up of one hidden layer containing 768 units and a tanh activation function followed by a sigmoid output. Between the RoBERTa model and its classifier, the first dimension of RoBERTa's output tensor (also annotated as the beginning of sentence token <s>) was extracted and treated as the input of the classifier. A dropout with a rate of 0.1 was added before the hidden layer to prevent overfitting. We utilized this classifier structure for all of our three methods and fine-tuned the whole model with officially recommended hyperparameters *(epochs: 2, batch size: 32, optimizer: Adam, learning rate: 1×10^{-5})*.

3.6 Baselines

Jiang and colleagues (2018; 2019) investigated and published a set of outstanding methods based on Word Embedding algorithms and the LSTM neural network, which outperformed those using human-engineered features with conventional classification models. Using a large corpus of unlabeled tweets, their approach generated a vector space model (VSM) to encode the words and trained and tested an LSTM-based classifier with a smaller set of annotated tweets. In our approach, we built the same (baseline) models by following the published structures and procedures: a VSM built by word2vec, GloVe and fastText algorithms with 128 dimensions and an LSTM layer with 128 hidden units and L2 regularizer followed by a fully connected layer with the sigmoid output. The models were trained by an Adam optimizer with a learning rate of 2×10^{-4} and a batch size of 32 for 5 epochs.

3.7 Data

Two corpora of Twitter data were used in our work.

A total of 22 million raw tweets were collected using Twitter Streaming APIs from August 25, 2015, to December 7, 2016, and another set of 52 million raw tweets was collected from 2006 to 2017 using a home-made crawler based upon the permission policy specified in Twitter's robots.txt file. Both sets were gathered by searching tweets with the keywords of a set of brand and generic medication names. These two corpora were merged and filtered. After dropping duplicates and eliminating non-English twitters, a corpus of 10 million tweets was collected. To study the changes in classification performance, the same corpus of 12,331 annotated tweets, published on Github by (Jiang, et al., 2018), was utilized.

For this task, the corpus of 10 million cleaned tweets were selected for training the Twitter RoBERTa model from scratch as well as updating the LM – note that the both LM updating and training from scratch procedures did not use any labels of the annotation and the annotated 12K tweets were excluded from the 10 million tweets. Interestingly, the baseline methods used the same 10 million raw tweets to build vector space models of neural embedding. Likewise, the baseline classifiers were also trained and tested with 12,331 labeled tweets. Table 1 lists the composition of annotated tweets.

	PETs	Non-PETs	Total
Tweet Count	2,962	9,369	12,331

Table 1. Composition of annotated tweets.

3.8 Statistical Analysis

To determine if any differences in the results among different methods could be due to chance, we conducted statistical analyses on the results between our methods and baseline methods. In our hypothesis testing, the null hypothesis was that the difference between a pair of method does not exist (null hypothesis) while the data remain the same. To do so, we partitioned data into the same subsets for all the methods in cross-validation – that is, each fold has the same set of tweets for different methods. This treatment facilitated us to use the paired t-test on the performance measures of each pair of the method. We set the p-value threshold to 0.05, meaning that any p-value less than 0.05 ($p < 0.05$) indicates that the difference does exist and it is not due to chance.

Method	Accuracy	Precision (PET)	Recall (PET)	F1 (PET)	AUC/ROC (PET)
Updated RoBERTa(160K)[1]	0.873	0.732	0.760	0.745	0.933
Updated RoBERTa(106K)[1]	**0.879**	**0.751**	0.746	0.748	**0.934**
Updated RoBERTa(53K)[1]	0.877	0.734	**0.775**	**0.754**	0.932
RoBERTa Original[2]	0.866	0.712	0.759	0.735	0.925
Twitter RoBERTa(160K)[3]	0.859	0.690	0.762	0.724	0.921
Twitter RoBERTa(106K)[3]	0.859	0.706	0.730	0.718	0.917
Twitter RoBERTa(53K)[3]	0.855	0.699	0.709	0.704	0.911
word2vec-LSTM	0.844	0.693	0.661	0.677	0.898
GloVe-LSTM	0.839	0.683	0.651	0.667	0.892
fastText-LSTM	0.842	0.681	0.663	0.672	0.891

1 Updated RoBERTa in 160k, 106k, 53k steps

2 Facebook's pre-trained RoBERTa

3 Twitter RoBERTa in 160k, 106k, 53k steps

Table 2. Classification performance. The last 3 rows are for baseline methods.

4   Results

To compare the performance differences between our methods and baseline methods, 10-fold cross-validation was conducted for each method and the mean value of each classification measure was collected. Table 2 shows the measures of the classification performance between our methods and baselines' (the highest values are in boldface).

Table 3 (in appendix) lists the statistical analysis results of each performance measure in cross-validation between our methods and baseline methods.

5   Discussions

According to the results in Table 2, we can see that compared to baseline methods, the approaches of RoBERTa model with or without updating achieved better performance in all measures, and the Twitter RoBERTa model trained with our data also performed better except in precision, and such differences were confirmed to exist statistically by the p-values in Table 3 ($p < 0.05$). In general, we can consider that the RoBERTa models performed better than Word Embedding + LSTM method in this task.

A noticeable improvement between pre-trained and updated RoBERTa models and baseline methods is the precision and recall, whereas the precision of Twitter RoBERTa model remained relatively unchanged at the same time. The recall is the sensitivity of how many true instances are predicted correctly and precision rates how many

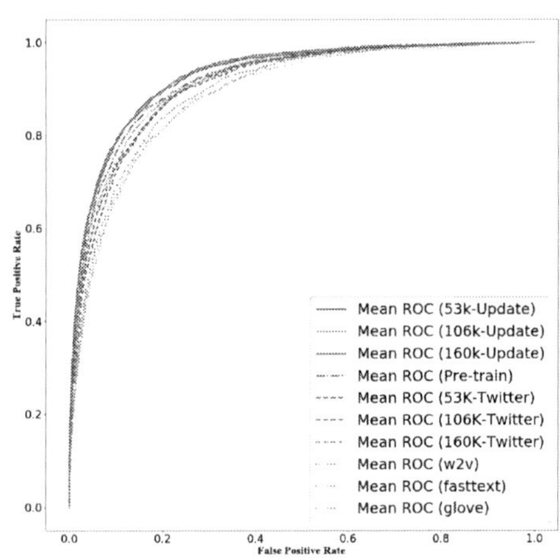

Figure 4. The ROC curves of our methods.

positive predictions are correct. A higher recall could help the model discover more potential positive instances and higher precision means more true positives (TP) and less false positives (FP) in the prediction. In other words, RoBERTa models can improve the sensitivity and identify PETs more precisely, resulting in more true positives in the predicted PET class.

Another remarkable measure could be the ROC/AUC score, which was also improved significantly as shown by the curves in Figure 4. ROC (Receiver Operating Characteristic) is a curve plotting true positive rate (TPR, or sensitivity) in the y-axis and false positive rate (FPR, or 1-specificity) in the x-axis, and is commonly used to show how well the model can distinguish two different objects. The area under the curve (AUC) of ROC is used to quantify the score of ROC. The results in Table 3 show that the lowest p-value between our methods and baseline methods is ROC, which may imply that ROC was improved most significantly among all performance measures. That is to say, our methods can be good choices with improved ROCs in this task and they are much more robust in distinguishing PETs and non-PETs.

Our methods also achieved a modest improvement in accuracy, but it could not be interpreted as that better accuracy leads to better performance. Because our dataset is imbalanced (PETs: non-PETs = 1: 3.16, as shown in Table 1) and accuracy is based upon the prediction of both positive and negative classes, higher accuracy could be attributed to the imbalance. Thus, accuracy is not an important measure that should be of concern.

The results also show that performing LM updating before classifier fine-tuning could yield more improvement in accuracy, precision, F1, and AUC. Nevertheless, the p-values indicate that they are not significant if updating the LM for more steps. But as for the Twitter RoBERTa model, which was trained from scratch, the steps of training affected performances in some measures which were supported by our statistical analysis. This outcome suggests that a larger number of steps are needed for performance improvement when training from scratch, and small steps are enough for LM updating to achieve better performance than the original RoBERTa model.

The possible reason for the improvement of these RoBERTa-based methods over baseline approaches could be attributed to the level of features. As is known, the features extracted by VSM such as word2vec, which is based upon word-level and co-occurrence. But RoBERTa, which extracts contextual-level features, maybe more powerful in processing tweet-like text which is poisoned by misspelling and incorrect grammars. The possible explanation for the performance difference between Updated RoBERTa and Twitter RoBERTa can be the slow learning process. The updating process is based on the pre-trained RoBERTa model, which is already pre-trained with a very large dataset by Facebook. It may be easier to adapt itself to our dataset, and the larger number of updating steps did less to help improve performance. But for Twitter RoBERTa, since it was trained from scratch and only 15% of

tokens were randomly masked, the model could only learn a small part of sentences for each step. Therefore, it may take more time to learn the data distribution, and the larger number of training steps is recommended.

6   Conclusion

In this study, we investigated different ways to use Facebook's RoBERTa model to improve performance in predicting personal experience tweets on medication use. Our results demonstrated that using the fine-tuning method on the pre-trained RoBERTa model achieved better classification performance than previous Word Embedding + LSTM methods, and the original pre-trained RoBERTa could perform better than training a new RoBERTa model from scratch. More importantly, updating the pre-trained RoBERTa language model with our data could yield better performance. The 10-fold cross-validation was used to test statistically the performance differences between our approaches and baseline methods. The results confirmed that the improvement does exist with statistical significance ($p < 0.05$). This suggests the pre-trained RoBERTa model and LM updating method are better choices for this task and significantly boost the capability to identify personal experience tweets. It is conceivable that our method could apply to other classification tasks using Twitter data related to health issues.

7   Acknowledgement

Authors wish to thank College of Technology at Purdue University Northwest for providing funding to support this work.

8   References

Alvaro, N., Conway, M., Doan, S., Lofi, C., Overington, J. and Collier, N., 2015. Crowdsourcing Twitter annotations to identify first-hand experiences of prescription drug use. Journal of biomedical informatics, 58, pp.280-287.

Aramaki, E., Maskawa, S. and Morita, M., 2011, July. Twitter catches the flu: detecting influenza epidemics using Twitter. In Proceedings of the conference on empirical methods in natural language processing (pp. 1568-1576). Association for Computational Linguistics.

Bian, J., Topaloglu, U. and Yu, F., 2012, October. Towards large-scale twitter mining for drug-related adverse events. In Proceedings of the 2012 international workshop on Smart health and wellbeing (pp. 25-32). ACM.

Bojanowski, P., Grave, E., Joulin, A. and Mikolov, T., 2017. Enriching word vectors with subword information. Transactions of the Association for Computational Linguistics, 5, pp.135-146.

Byrd, K., Mansurov, A. and Baysal, O., 2016, May. Mining Twitter data for influenza detection and surveillance. In Proceedings of the International Workshop on Software Engineering in Healthcare Systems (pp. 43-49). ACM.

Calix, R.A., Gupta, R., Gupta, M. and Jiang, K., 2017, November. Deep gramulator: Improving precision in the classification of personal health-experience tweets with deep learning. In 2017 IEEE International Conference on Bioinformatics and Biomedicine (BIBM) (pp. 1154-1159). IEEE.

Coppersmith, G., Dredze, M. and Harman, C., 2014, June. Quantifying mental health signals in Twitter. In Proceedings of the workshop on computational linguistics and clinical psychology: From linguistic signal to clinical reality (pp. 51-60).

Devlin, J., Chang, M.W., Lee, K. and Toutanova, K., 2018. Bert: Pre-training of deep bidirectional transformers for language understanding. arXiv preprint arXiv:1810.04805.

Gesualdo, F., Stilo, G., Gonfiantini, M.V., Pandolfi, E., Ve-lardi, P. and Tozzi, A.E., 2013. Influenza-like illness surveil-lance on Twitter through automated learning of naïve language. PLoS One, 8(12), p.e82489.

Heaivilin, N., Gerbert, B., Page, J.E. and Gibbs, J.L., 2011. Public health surveillance of dental pain via Twitter. Journal of dental research, 90(9), pp.1047-1051.

Ji, X., Chun, S.A. and Geller, J., 2012, April. Epidemic outbreak and spread detection system based on twitter data. In International Conference on Health Information Science (pp. 152-163). Springer, Berlin, Heidelberg.

Jiang, K., Calix, R. and Gupta, M., 2016, August. Construction of a personal experience tweet corpus for health surveillance. In Proceedings of the 15th workshop on biomedical natural language processing (pp. 128-135).

Jiang, K., Feng, S., Calix, R.A. and Bernard, G.R., 2019, January. Assessment of word embedding techniques for identification of personal experience tweets pertaining to medication uses. In International Workshop on Health Intelligence (pp. 45-55). Springer, Cham.

Jiang, K., Feng, S., Song, Q., Calix, R.A., Gupta, M.

and Bernard, G.R., 2018. Identifying tweets of personal health experience through word embedding and LSTM neural network. BMC bioinformatics, 19(8), p.210.

Jiang, K. and Zheng, Y., 2013, December. Mining twitter data for potential drug effects. In International conference on advanced data mining and applications (pp. 434-443). Springer, Berlin, Heidelberg.

Kagashe, I., Yan, Z. and Suheryani, I., 2017. Enhancing seasonal influenza surveillance: topic analysis of widely used medicinal drugs using Twitter data. Journal of medical Internet research, 19(9), p.e315.

Lee, K., Agrawal, A. and Choudhary, A., 2013, August. Real-time disease surveillance using twitter data: demonstration on flu and cancer. In Proceedings of the 19th ACM SIGKDD international conference on Knowledge discovery and data mining (pp. 1474-1477). ACM.

Liu, Y., Ott, M., Goyal, N., Du, J., Joshi, M., Chen, D., Levy, O., Lewis, M., Zettlemoyer, L. and Stoyanov, V., 2019. Roberta: A robustly optimized bert pretraining approach. arXiv preprint arXiv:1907.11692.

Mikolov, T., Chen, K., Corrado, G. and Dean, J., 2013. Efficient estimation of word representations in vector space. arXiv preprint arXiv:1301.3781.

Parker, J., Wei, Y., Yates, A., Frieder, O. and Goharian, N., 2013, August. A framework for detecting public health trends with twitter. In Proceedings of the 2013 IEEE/ACM International Conference on Advances in Social Networks Analysis and Mining (pp. 556-563). ACM.

Paul, M.J. and Dredze, M., 2011, July. You are what you tweet: Analyzing twitter for public health. In Fifth International AAAI Conference on Weblogs and Social Media.

Paul, M.J., Dredze, M., Broniatowski, D.A. and Generous, N., 2015, April. Worldwide influenza surveillance through twitter. In Workshops at the Twenty-Ninth AAAI Conference on Artificial Intelligence.

Pennington, J., Socher, R. and Manning, C., 2014, October. Glove: Global vectors for word representation. In Proceedings of the 2014 conference on empirical methods in natural language processing (EMNLP) (pp. 1532-1543).

Reece, A.G., Reagan, A.J., Lix, K.L., Dodds, P.S., Danforth, C.M. and Langer, E.J., 2017. Forecasting the onset and course of mental illness with Twitter data. Scientific reports, 7(1), p.13006.

Sennrich, R., Haddow, B. and Birch, A., 2015. Neural

machine translation of rare words with subword units. arXiv preprint arXiv:1508.07909.

Thackeray, R., Burton, S.H., Giraud-Carrier, C., Rollins, S. and Draper, C.R., 2013. Using Twitter for breast cancer prevention: an analysis of breast cancer awareness month. BMC cancer, 13(1), p.508.

Wijeratne, S., Sheth, A., Bhatt, S., Balasuriya, L., Al-Olimat, H. S., Gaur, M., Yazdavar, A. H., Thirunarayan, K.: Feature Engineering for Twitter-based Applications. Feature Engineering for Machine Learning and Data Analytics, 35 (2017).

A Appendices

Method	Measure	RoBERTa	RoBERTa-Updated(53K)	RoBERTa-Updated(106K)	RoBERTa-Updated(160K)	RoBERTa-Twitter(53K)	RoBERTa-Twitter(106K)	RoBERTa-Twitter(160K)
RoBERTa	Acc.		5.106×10^{-2}	**7.908×10^{-4}**	1.036×10^{-1}	**2.153×10^{-2}**	**2.595×10^{-2}**	7.047×10^{-2}
	Prec.		1.793×10^{-1}	**3.950×10^{-3}**	2.075×10^{-1}	2.344×10^{-1}	3.248×10^{-1}	1.071×10^{-1}
	Recall		2.170×10^{-1}	1.651×10^{-1}	4.853×10^{-1}	**4.586×10^{-2}**	1.096×10^{-1}	4.484×10^{-1}
	F1		**8.196×10^{-4}**	**1.247×10^{-3}**	**2.306×10^{-2}**	**5.759×10^{-3}**	**1.576×10^{-2}**	9.804×10^{-2}
	AUC		**1.312×10^{-4}**	**7.787×10^{-5}**	**2.650×10^{-4}**	**2.397×10^{-5}**	**6.578×10^{-4}**	**3.026×10^{-2}**
RoBERTa-Updated(53K)	Acc.	5.106×10^{-2}		3.204×10^{-1}	2.880×10^{-1}	**6.821×10^{-4}**	**4.198×10^{-3}**	**9.187×10^{-4}**
	Prec.	1.793×10^{-1}		1.774×10^{-1}	4.660×10^{-1}	**3.713×10^{-2}**	1.694×10^{-1}	**1.699×10^{-2}**
	Recall	2.170×10^{-1}		**4.363×10^{-2}**	2.656×10^{-1}	**1.410×10^{-2}**	9.040×10^{-2}	3.019×10^{-1}
	F1	**8.196×10^{-4}**		1.094×10^{-1}	**4.049×10^{-2}**	**1.948×10^{-4}**	**1.954×10^{-6}**	**6.061×10^{-5}**
	AUC	**1.312×10^{-4}**		8.087×10^{-2}	5.282×10^{-2}	**9.169×10^{-8}**	**2.980×10^{-7}**	**7.746×10^{-6}**
RoBERTa-Updated(106K)	Acc.	**7.908×10^{-4}**	3.204×10^{-1}		7.762×10^{-2}	**4.148×10^{-5}**	**1.415×10^{-5}**	**6.621×10^{-5}**
	Prec.	**3.950×10^{-3}**	1.774×10^{-1}		1.682×10^{-1}	**5.194×10^{-3}**	**1.010×10^{-2}**	**9.835×10^{-4}**
	Recall	1.651×10^{-1}	**4.363×10^{-2}**		2.882×10^{-1}	1.199×10^{-1}	2.575×10^{-1}	2.289×10^{-1}
	F1	**1.247×10^{-3}**	1.094×10^{-1}		1.449×10^{-1}	**3.382×10^{-4}**	**1.216×10^{-4}**	**2.713×10^{-4}**
	AUC	**7.787×10^{-5}**	8.087×10^{-2}		3.416×10^{-1}	**1.357×10^{-7}**	**5.519×10^{-7}**	**8.113×10^{-6}**
RoBERTa-Updated(160K)	Acc.	1.036×10^{-1}	2.880×10^{-1}	7.762×10^{-2}		**3.212×10^{-3}**	**3.471×10^{-3}**	**4.285×10^{-3}**
	Prec.	2.075×10^{-1}	4.660×10^{-1}	1.682×10^{-1}		6.953×10^{-2}	1.398×10^{-1}	**2.216×10^{-2}**
	Recall	4.853×10^{-1}	2.656×10^{-1}	2.882×10^{-1}		**3.748×10^{-2}**	1.779×10^{-1}	4.702×10^{-1}
	F1	**2.306×10^{-2}**	**4.049×10^{-2}**	1.449×10^{-1}		**1.828×10^{-4}**	**1.204×10^{-3}**	**6.101×10^{-3}**
	AUC	**2.650×10^{-4}**	5.282×10^{-2}	3.416×10^{-1}		**8.982×10^{-9}**	**6.030×10^{-8}**	**7.682×10^{-7}**
RoBERTa-Twitter(53K)	Acc.	**2.153×10^{-2}**	**6.821×10^{-4}**	**4.148×10^{-5}**	**3.212×10^{-3}**		2.124×10^{-1}	1.908×10^{-1}
	Prec.	2.344×10^{-1}	**3.713×10^{-2}**	**5.194×10^{-3}**	6.953×10^{-2}		3.888×10^{-1}	2.640×10^{-1}
	Recall	**4.586×10^{-2}**	**1.410×10^{-2}**	1.199×10^{-1}	**3.748×10^{-2}**		2.817×10^{-1}	**2.878×10^{-2}**
	F1	**5.759×10^{-3}**	**1.948×10^{-4}**	**3.382×10^{-4}**	**1.828×10^{-4}**		1.791×10^{-1}	**2.271×10^{-2}**
	AUC	**2.397×10^{-5}**	**9.169×10^{-8}**	**1.357×10^{-7}**	**8.982×10^{-9}**		**1.873×10^{-3}**	**2.129×10^{-5}**
RoBERTa-Twitter(106K)	Acc.	**2.595×10^{-2}**	**4.198×10^{-3}**	**1.415×10^{-5}**	**3.471×10^{-3}**	2.124×10^{-1}		4.596×10^{-1}
	Prec.	3.248×10^{-1}	1.694×10^{-1}	**1.010×10^{-2}**	1.398×10^{-1}	3.888×10^{-1}		1.801×10^{-1}
	Recall	1.096×10^{-1}	9.040×10^{-2}	2.575×10^{-1}	1.779×10^{-1}	2.817×10^{-1}		8.020×10^{-2}
	F1	**1.576×10^{-2}**	**1.954×10^{-6}**	**1.216×10^{-4}**	**1.204×10^{-3}**	1.791×10^{-1}		**4.668×10^{-2}**
	AUC	**6.578×10^{-4}**	**2.980×10^{-7}**	**5.519×10^{-7}**	**6.030×10^{-8}**	**1.873×10^{-3}**		**1.630×10^{-3}**
RoBERTa-Twitter(160K)	Acc.	7.047×10^{-2}	**9.187×10^{-4}**	**6.621×10^{-5}**	**4.285×10^{-3}**	1.908×10^{-1}	4.596×10^{-1}	
	Prec.	1.071×10^{-1}	**1.699×10^{-2}**	**9.835×10^{-4}**	**2.216×10^{-2}**	2.640×10^{-1}	1.801×10^{-1}	
	Recall	4.484×10^{-1}	3.019×10^{-1}	2.289×10^{-1}	4.702×10^{-1}	**2.878×10^{-2}**	8.020×10^{-2}	
	F1	9.804×10^{-2}	**6.061×10^{-5}**	**2.713×10^{-4}**	**6.101×10^{-3}**	**2.271×10^{-2}**	**4.668×10^{-2}**	
	AUC	**3.026×10^{-2}**	**7.746×10^{-6}**	**8.113×10^{-6}**	**7.682×10^{-7}**	**2.129×10^{-5}**	**1.630×10^{-3}**	

Table 3a. Statistical analysis results (p values) for RoBERTa models. Values in boldface are less than 0.05.

Method	Measure	RoBERTa	RoBERTa-Updated(53K)	RoBERTa-Updated(106K)	RoBERTa-Updated(160K)	RoBERTa-Twitter(53K)	RoBERTa-Twitter(106K)	RoBERTa-Twitter(160K)
Word2Vec-LSTM	Acc.	1.663×10^{-3}	1.879×10^{-5}	1.919×10^{-6}	1.050×10^{-4}	7.198×10^{-3}	5.770×10^{-3}	4.898×10^{-3}
	Prec.	1.861×10^{-1}	4.768×10^{-2}	6.659×10^{-3}	9.901×10^{-2}	4.045×10^{-1}	2.818×10^{-1}	4.422×10^{-1}
	Recall	2.603×10^{-2}	1.062×10^{-2}	3.799×10^{-2}	2.877×10^{-2}	1.864×10^{-1}	5.933×10^{-2}	2.029×10^{-2}
	F1	1.145×10^{-2}	1.878×10^{-3}	2.958×10^{-3}	3.891×10^{-3}	7.551×10^{-2}	2.553×10^{-2}	1.450×10^{-2}
	AUC	2.582×10^{-7}	8.251×10^{-8}	5.581×10^{-8}	4.686×10^{-8}	1.479×10^{-5}	1.864×10^{-5}	5.317×10^{-7}
Glove-LSTM	Acc.	9.613×10^{-5}	8.448×10^{-5}	1.213×10^{-5}	2.637×10^{-4}	1.236×10^{-2}	1.137×10^{-3}	5.019×10^{-4}
	Prec.	1.005×10^{-1}	2.234×10^{-2}	3.395×10^{-3}	2.055×10^{-2}	2.617×10^{-1}	1.686×10^{-1}	3.822×10^{-1}
	Recall	1.442×10^{-2}	3.515×10^{-3}	1.768×10^{-2}	1.818×10^{-3}	1.008×10^{-1}	4.343×10^{-2}	1.303×10^{-2}
	F1	1.155×10^{-4}	1.451×10^{-5}	2.706×10^{-5}	1.673×10^{-5}	4.342×10^{-3}	1.953×10^{-3}	7.256×10^{-4}
	AUC	7.183×10^{-9}	1.086×10^{-9}	3.358×10^{-10}	1.338×10^{-10}	1.994×10^{-9}	1.326×10^{-8}	2.239×10^{-9}
Fasttext-LSTM	Acc.	9.961×10^{-5}	5.171×10^{-5}	1.029×10^{-6}	2.716×10^{-4}	7.583×10^{-3}	2.108×10^{-3}	7.676×10^{-3}
	Prec.	3.035×10^{-2}	1.449×10^{-2}	1.133×10^{-4}	2.920×10^{-2}	1.864×10^{-1}	1.425×10^{-1}	3.588×10^{-1}
	Recall	2.448×10^{-5}	1.183×10^{-4}	4.201×10^{-5}	9.241×10^{-4}	9.009×10^{-2}	1.946×10^{-2}	3.609×10^{-3}
	F1	8.453×10^{-8}	6.394×10^{-9}	3.063×10^{-8}	9.138×10^{-8}	1.282×10^{-3}	1.064×10^{-4}	5.055×10^{-6}
	AUC	1.011×10^{-8}	3.002×10^{-9}	1.344×10^{-8}	3.410×10^{-9}	9.257×10^{-8}	2.562×10^{-7}	1.066×10^{-8}

Table 3b. Statistical analysis results (p values) for baselines. Values in boldface are less than 0.05.

Detecting Foodborne Illness Complaints in Multiple Languages Using English Annotations Only

Ziyi Liu, Giannis Karamanolakis, Daniel Hsu, Luis Gravano
Columbia University, New York, NY 10027, USA
`zl2888@columbia.edu,{gkaraman,djhsu,gravano}@cs.columbia.edu`

Abstract

Health departments have been deploying text classification systems for the early detection of foodborne illness complaints in social media documents such as Yelp restaurant reviews. Current systems have been successfully applied for documents in English and, as a result, a promising direction is to increase coverage and recall by considering documents in additional languages, such as Spanish or Chinese. Training previous systems for more languages, however, would be expensive, as it would require the manual annotation of many documents for each new target language. To address this challenge, we consider cross-lingual learning and train multilingual classifiers using only the annotations for English-language reviews. Recent zero-shot approaches based on pre-trained multi-lingual BERT (mBERT) have been shown to effectively align languages for aspects such as sentiment. Interestingly, we show that those approaches are less effective for capturing the nuances of foodborne illness, our public health application of interest. To improve performance without extra annotations, we create artificial training documents in the target language through machine translation and train mBERT jointly for the source (English) and target language. Furthermore, we show that translating labeled documents to multiple languages leads to additional performance improvements for some target languages. We demonstrate the benefits of our approach through extensive experiments with Yelp restaurant reviews in seven languages. Our classifiers identify foodborne illness complaints in multilingual reviews from the Yelp Challenge dataset, which highlights the potential of our general approach for deployment in health departments.

1 Introduction

With the rise of social media, more and more users post online documents where they disclose serious

Wahoo's Fish Taco- Las Vegas ⊘ Claimed

☆☆☆☆☆ 1/14/2017

I recently went to this location and ordered the chicken rice bowl. Later that night I started to feel not so good. This was the only thing I had eaten that day so I know food poisoning when it happens. I spoke to a friend of mine who had chicken tacos and he also told he had gotten food poisoning also. I would say stay clear from the chicken !!

Basha – Sherbrooke ⊘ Unclaimed

☆☆☆☆☆ 4/4/2018

千！万！别！去！我男朋友昨天晚上点了个shawarma plate，从凌晨三点开始上吐下泻到现在。我认识他五年，连感冒都没见他得过。珍爱生命远离这家餐馆吧。

La Mojarra Loca Grill ⊘ Unclaimed

☆☆☆☆☆ 7/23/2017

Este lugar la verdad no se los recomiendo y más si se trata para los niños. Fui con mi familia al lunch y mi niño pidió chicken nuggets y de verdad se los digo esos pedazos de pollo estaban asquerosos parece que los tenían de hace mucho tiempo y el de inmediato empezó a vomitar es increíble que un niño de 4 años te diga que la comida no sirve eso para el chef. ...

Figure 1: Examples of Yelp restaurant reviews discussing food poisoning in different languages.

incidents, such as getting food poisoning from a restaurant. As many of those incidents may not be reported through established complaint systems, health departments have deployed text classification systems for the identification of social media documents, such as Yelp reviews and tweets, that discuss foodborne illness episodes. Figure 1 shows examples of Yelp restaurant reviews discussing food poisoning in English, Chinese, and Spanish.

Current classification systems have been applied for documents written in English and deployed in several health departments, including those in Chicago (Harris et al., 2014), Nevada (Sadilek et al., 2016), New York City (Effland et al., 2018), and St. Louis (Harris et al., 2018). Online documents flagged by the classifiers are typically analyzed by epidemiologists, who further investigate the incidents (e.g., by inspecting the corresponding restau-

Proceedings of the 11th International Workshop on Health Text Mining and Information Analysis, pages 138–146
November 20, 2020. ©2020 Association for Computational Linguistics
https://doi.org/10.18653/v1/P17

rants). This process contributes to the early detection of previously unknown foodborne outbreaks. Given the success of current systems, a promising new direction is to extend these systems to use non-English languages, thus increasing their coverage and capacity to identify foodborne outbreaks.

Directly applying existing techniques for foodborne illness detection to other languages would be expensive and time-consuming. Current (supervised) classifiers have been trained on thousands of documents that were manually labeled with binary ("Sick" vs. "Not Sick") labels provided by epidemiologists, and it would be expensive to replicate this effort for new target languages. Furthermore, it is hard to collect documents for annotation for our task because most online documents do not discuss foodborne illness. Alternative approaches beyond supervised learning are thus required to efficiently scale to multiple languages.

To address the costly requirement of supervised learning approaches, we train multilingual classifiers through a less expensive *cross-lingual* text classification approach. For a given non-English target language, our approach does not require manually annotated in-language documents but instead trains classifiers using the already available English annotations. We follow recent techniques for cross-lingual text classification and employ pre-trained multi-lingual BERT (mBERT) representations (Wu and Dredze, 2019; Pires et al., 2019). However, while pre-trained mBERT representations have been shown to be effective for tasks such as cross-lingual sentiment classification (Wu and Dredze, 2019), we show that such representations are less effective for capturing the nuances of foodborne illness, which is required by our application of focus. To improve performance, we translate labeled English reviews to the target language and fine-tune mBERT *jointly* for both languages, which turns out to be more effective than fine-tuning on either language separately. Furthermore, we show that fine-tuning mBERT for multiple languages in parallel leads to additional improvements for some target languages such as German and Italian.

Our work makes the following contributions:

1. We present a cross-lingual learning approach for foodborne illness detection in non-English social media documents. Our approach is efficient and requires only English labeled data.

2. We show how to improve the performance of pre-trained mBERT for our rare classification task. Our preliminary results show that generating additional artificial training data in multiple languages through machine translation leads to promising improvements over zero-shot mBERT.

3. We evaluate our approach on Yelp reviews in English, Spanish, Chinese, French, German, Japanese, and Italian. Our approach substantially outperforms previous techniques and baselines for this task. Our multilingual classifiers successfully identify foodborne illness across languages in reviews from the Yelp Challenge dataset, which highlights the potential of our approach for successful, real-world deployment in health departments.

The rest of this paper is organized as follows. In Section 2, we provide the necessary background for our work. In Section 3, we describe our approach for cross-lingual foodborne detection. In Section 4, we present the experimental setup and results. In Section 5, we conclude and suggest future work.

2 Background

In this section, we provide background on foodborne illness detection (Section 2.1) and cross-lingual text classification (Section 2.2).

2.1 Foodborne Illness Detection in English Documents

Foodborne illness detection in online documents has been addressed as a binary text classification task: the goal is to train a classifier that, given the text of a document, predicts a binary ("Sick" vs. "Not Sick") label, corresponding to whether the document is mentioning foodborne illness or not. Sadilek et al. (2016) trained support vector machine classifiers (based on unigram, bigram and trigram features) using 8,000 tweets that were independently labeled by five human annotators. Effland et al. (2018) trained classifiers using more than 10,000 Yelp reviews that were manually annotated by epidemiologists. The paper compares several methods and found that logistic regression had the best performance. Karamanolakis et al. (2019) trained a weakly-supervised neural network that predicts a label for each individual sentence of a review and improves the recall of foodborne illness complaints compared to the best performing classifier in Effland et al. (2018).

2.2 Cross-Lingual Text Classification

Cross-lingual text classification trains a classifier on a target language T by leveraging labeled documents in a source language S. We focus on the challenging cross-lingual classification setting where only unlabeled documents are available in T.

Some effective approaches address cross-lingual classification by relying on cross-lingual word embeddings (Gouws and Søgaard, 2015; Ruder et al., 2019), which represent words from different languages in the same vector space, where words across languages with similar meanings are represented as similar vectors. Cross-lingual word embeddings facilitate cross-lingual model transfer as a classifier trained on labeled documents in S could be directly applied for test documents in T.

More recent approaches addressed cross-lingual transfer using Multilingual BERT (Wu and Dredze, 2019; Pires et al., 2019; Karthikeyan et al., 2019; Rogers et al., 2020). Multilingual BERT, or mBERT, is a version of BERT (Devlin et al., 2019) that was trained on 104 languages in parallel. Training mBERT on English documents was shown to achieve impressively high performance on different target languages for several document classification tasks such as sentiment classification or topic detection (Rogers et al., 2020). The successful application of mBERT for various cross-lingual tasks inspired us to employ mBERT for our public-health application, as we describe next.

3 Foodborne Illness Detection in Multiple Languages

We now define our problem of focus (Section 3.1) and describe our cross-lingual learning approach (Sections 3.2 and 3.3).

3.1 Problem Definition

Our goal is to address foodborne illness detection in non-English languages where labeled documents are not available. As the collection of manual annotations for each new language is an expensive and time-consuming proposition, we focus on training multilingual classifiers using only already available English documents. More formally, we assume access to a source language S (English) with a labeled dataset $D_S = \{(x_i^S, y_i^S)\}$, where x_i^S is a source language document and y_i^S is the corresponding binary ("Sick" vs. "Not Sick") label. For a target language T we assume access to a dataset D_T of unlabeled target documents x^T. Our goal is

to train a classifier for the target language T that, given an unseen test document x^T in T, predicts a binary ("Sick" vs. "Not Sick") label.

3.2 Fine-Tuning mBERT on S and T

To address the task mentioned in Section 3.1, we use pre-trained mBERT representations, which effectively align representations of different languages (Section 2.2).

It has been shown that mBERT achieves impressive zero-shot performance for tasks such as sentiment classification and topic detection (Wu and Dredze, 2019; Pires et al., 2019): fine-tuning mBERT on the labeled dataset D_S in S leads to accurate classification of unlabeled documents x^T in T, possibly because representations across languages are well aligned with respect to the target sentiment or topic. However, in contrast to previous tasks, we show that zero-shot mBERT is not effective for foodborne detection. We hypothesize that this discrepancy is observed because pre-trained mBERT representations are not effectively aligned across languages with respect to the aspect of foodborne illness, which may be rarely mentioned in documents used for pre-training mBERT.

To address this issue and improve classification performance for our task, we do not consider zero-shot training but fine-tune mBERT in both S and T. Our main idea is that fine-tuning mBERT in documents from both S and T will encourage a stronger alignment of the cross-lingual representations with respect to the aspect of foodborne illness. The main challenge associated with our approach is that labeled documents are not available in the target language T.

To generate training documents in T, we translate labeled documents x_i^S from S (English) to T using machine translation. In particular, we assume that machine translation is sufficiently accurate to the extent that the translated document $x_i^{S \to T}$ has the same label as the original document x_i^S. Under this assumption, we generate a weakly annotated dataset $D_T' = \{(x_i^{S \to T}, y_i^S)\}$ by translating all documents x_i^S annotated as "Sick" and an equal number of documents randomly sampled from "Not Sick" documents in D_S. Then, we increase the size of D_T' by sampling unlabeled documents x_i^T from D_T uniformly at random. Each sampled document is assigned the "Not Sick" label as the chance of randomly choosing a document mentioning foodborne illness is very low. The number of sampled

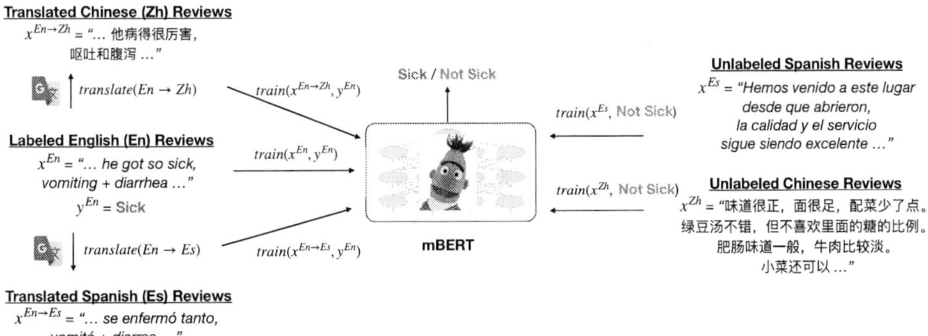

Figure 2: Our training procedure. We translate *labeled* English reviews to the target languages and use the translated reviews with the original labels as extra training samples. We also use a sample of *unlabeled* multilingual reviews as negative ("Not Sick") training examples.

documents is chosen so that the total number of "Not Sick" documents in D'_T is equal to that in D_S.

After creating the weakly labeled D'_T set we fine-tune our mBERT-based classifier jointly on D_S and D_T by concatenating and shuffling the two datasets. As we will show, this training procedure is more effective than fine-tuning mBERT on D_S or D_T separately.

3.3 Considering Multiple Source Languages

Classification performance in T may potentially improve using *multiple* source languages $\{S_1, \ldots, S_K\}$ other than S (English) for which unlabeled documents and machine translation systems are available. The main idea behind this approach is that training signals from multiple source languages could prevent overfitting to a single source language and as a result encourage mBERT to learn better cross-lingual representations for our task. Therefore, we adapt the procedure described in Section 3.2 to consider more source languages in addition to S and T, as we describe next.

To train mBERT using multiple source languages $S, S_1, \ldots, S_K$, we create a big training set that considers all source-language documents. In particular, first we create a weakly-labeled dataset D_{S_k} for each source language using machine translation, as we described in Section 3.2 for creating D_S. Then, we concatenate all source datasets $D_S, D'_{S_1}, \ldots, D'_{S_K}$ and fine-tune mBERT across all languages ($S, S_1, \ldots, S_K, T$). Note that, in our preliminary experiments, we have treated all languages as equal but in the future it would be interesting to consider alternative approaches, such as using different weights for examples from different languages. Figure 2 shows our overall training

procedure using English, Spanish and Chinese for training mBERT.

An important advantage of this approach is that the same mBERT classifier can be applied on any target language T supported in mBERT. As a result, deployment in health departments would be easier since it involves a single model for all languages and does not require extra pre-processing steps such as running a language detector[1] for each test document and applying language-specific models. Also, as we will show next, considering multiple source languages during training encourages better generalization to a new *unseen* test language.

4 Experiments

We evaluate our approach on foodborne detection in English (En), Spanish (Es), Chinese (Zh), French (Fr), German (De), Japanese (Ja), and Italian (It).

4.1 Experimental Settings

Datasets. We use the same corpus of labeled English reviews from Effland et al. (2018). This dataset contains English reviews with ground truth annotations provided by epidemiologists. Table 1 reports the number of reviews on the train and test set. For details, see Effland et al. (2018).

We collect unlabeled multilingual reviews from Yelp restaurants in New York City (NYC), Los Angeles (LA), as well as other metropolitan areas in the Yelp Challenge dataset.[2] As the language of the

---

[1] In our experiments, language detectors sometimes predicted the wrong language for the text of a test restaurant review, for example because of multiple mentions of Italian dishes in a non-Italian review.

[2] https://www.kaggle.com/yelp-dataset/yelp-dataset

	All Reviews	**Sick**	**Not Sick**
Train	21,551	5894	15,657
Validation	1500	1090	410
Test	2975	949	2026

Table 1: Number of Yelp reviews in the English dataset with ground-truth (Sick vs. Not Sick) annotations.

	NYC Area	**LA Area**	**Yelp Challenge**	**Total**
Spanish	6267	11,458	2658	20,383
Chinese	1624	1488	603	3715
French	3882	741	24,807	29,430
German	2912	657	1394	4963
Japanese	2161	1469	563	4193
Italian	1259	322	173	1754

Table 2: Number of unlabeled Yelp reviews from the New York City area, Los Angeles area, as well as other metropolitan areas in the Yelp Challenge dataset.

reviews is not mentioned in the metadata, we used Python's langdetect[3] library to automatically detect the language. For evaluation on non-English languages, we translate the 2975 English test reviews to the target languages using the Google Translate API.[4]

Model Comparison. We compare the following models for our task:

- **Monolingual LogReg**: the logistic regression classifier that achieved the best results in (Effland et al., 2018). We train LogReg for a non-English target language T by translating English reviews to T using Google Translate (see Section 3.2).

- **Monolingual BERT**: a monolingual BERT classifier. Similarly to LogReg, we train BERT for a non-English target language T by translating English reviews to T using Google Translate.

- **mBERT**: a multilingual BERT classifier. We train mBERT on several combinations of languages using our approach described in Section 3.

Model Configuration. For LogReg, we tokenize text using Spacy[5] and convert the text documents to TF-IDF vectors.

For monolingual BERT, we consider pre-trained BERT representations from huggingface[6]:

- English: bert-base-uncased

- Spanish: dccuchile/bert-base-spanish-wwm-cased

- Chinese: bert-base-chinese

- French: camembert-base

- German: bert-base-german-cased

- Japanese: cl-tohoku/bert-base-japanese

- Italian: dbmdz/bert-base-italian-xxl-cased

For mBERT, we consider pre-trained mBERT representations from huffingface: bert-base-multilingual-cased. We fine-tuned BERT and mBERT using the Python simpletransformers[7] library. We did a hyperparameter search with BERT on English data using the validation set. The best hyperparameters are a learning rate of 1e-05, a batch size of 512, and a maximum sequence length of 512. We fine-tune BERT/mBERT for up to 5 epochs with early stopping based on the validation loss.

Evaluation Procedure. For each model, we choose the best set of hyperparameters according to the F1 score on the validation set. We report the following classification metrics on the test set: accuracy (Acc), precision (Prec), recall (Rec), and macro-average F1 score (F1).

4.2 Experimental Results

Table 3 shows F1 scores on all languages for various methods.

Monolingual BERT outperforms previous systems. Monolingual BERT outperforms LogReg: leveraging pre-trained contextual representations captures foodborne illness effectively.

Monolingual BERT outperforms mBERT. Interestingly, monolingual BERT performs better than mBERT. We hypothesize that, by focusing on a single language, pre-trained monolingual BERT representations capture foodborne-related aspects more effectively than mBERT representations that were pre-trained for all languages in parallel.

---

[3] https://pypi.org/project/langdetect/
[4] The Google Translate API was used in February 2020.
[5] https://spacy.io/api/tokenizer

[6] https://huggingface.co
[7] https://github.com/ThilinaRajapakse/simpletransformers

Model	Train Language	Test Language							AVG F1
		En	Es	Zh	Fr	De	Ja	It	
Monolingual LogReg	T	83.7	83.6	83.3	84.9	80.4	81.7	83.6	83.0
Monolingual BERT	T	91.6	91.3	**87.3**	92.4	**88.9**	**87.0**	90.7	89.9
mBERT	En	89.0	82.0	78.8	80.6	59.0	65.5	67.5	74.6
mBERT	T	89.0	87.1	87.0	88.6	87.3	88.8	**89.4**	88.2
mBERT	En $+ T$	89.0	**89.8**	89.7	90.5	88.0	89.4	88.2	89.2
mBERT	ALL	**91.3**	89.6	88.0	**90.7**	89.5	86.8	89.2	**89.3**

Table 3: F1 scores for various approaches evaluated on different test languages. Monolingual LogReg and BERT are trained on the translated documents in the target language T. mBERT is trained with various language configurations. Training mBERT in English and T is more effective than training on either language separately. Training mBERT across all 7 languages ("ALL") leads to further improvements for En, Fr, and De. Results in red correspond to the best performance across all models.

Model	Train	Es	Zh	AVG
mBERT	En	82.0	78.8	80.4
mBERT	ALL-T	**84.7**	**84.0**	**84.4**

Table 4: Zero-shot performance under two different settings: training on English-only data (En) vs. training on all languages except the target language (ALL-T). The latter approach performs substantially better than the former.

Model	Train	Acc	Prec	Rec	F1
LogReg	En	88.1	74.1	96.2	83.7
BERT	En	**94.4**	88.1	**95.4**	**91.6**
mBERT	En	92.5	83.8	95.0	89.0
mBERT	ALL	94.3	**89.2**	93.6	91.3

Table 5: Evaluation on English Yelp reviews.

Zero-shot mBERT is not effective. Training zero-shot mBERT using only English training data (En) is not effective and performs substantially worse than monolingual LogReg. This result validates our argument that pre-trained mBERT representations do not effectively capture the aspect of food poisoning, which is rarely mentioned in documents used for pre-training mBERT.

Artificial training reviews in T improve mBERT's performance. Translating English reviews to T and using translated reviews to train mBERT on T is substantially better than zero-shot mBERT trained on English directly. This result highlights the importance of in-language training documents, even if those documents are artificially created. Furthermore, training mBERT jointly on English and the target language T leads to better performance compared to training on each language separately.

Training on all languages leads to the best performance for mBERT. On average across languages, mBERT trained on all languages jointly performs better than other mBERT configurations with a single source language, but comparably to mBERT trained on En and T. Interestingly, for Chinese (Zh) and Japanese (Ja) performance is worse if more languages are added to the training set, possibly because these languages are more distant from Romance languages such as Spanish or French, and as a result considering those languages in the training set is not helpful.

Using multiple source languages leads to higher zero-shot performance. Table 4 shows results for the setting where we assume that documents from the target language are not available for training. Crucially, training mBERT on all languages except this target language performs substantially better than training mBERT only on English data, validating the importance of training mBERT on multiple languages jointly. Also, F1 scores when ignoring those languages during training (ALL-T) are lower by about 5 absolute points compared to considering them during training (ALL): we could potentially apply our approach to any unseen language out of the 104 languages that are supported by mBERT.

Detailed English results. Table 5 shows results in English. BERT (monolingual) has the best F1 score. Training mBERT on all languages (En, Es, Zh, Fr, De, Ja, It) is more effective than training mBERT on English-only labeled data. This validates our hypothesis that, by considering all languages, mBERT generalizes better to test reviews.

Model	Train	Acc	Prec	Rec	F1
LogReg	Es	87.9	73.7	**96.4**	83.6
LogReg*	En	88.2	75.5	93.4	83.5
BERT	Es	**94.2**	87.1	96.0	**91.3**
BERT*	En	93.6	87.1	93.9	90.4
mBERT	En	89.7	**92.3**	73.8	82.0
mBERT	Es	90.9	79.5	**96.4**	87.1
mBERT	En+Es	93.2	85.9	94.1	89.8
mBERT	ALL	93.3	89.0	90.2	89.6

(a) Results on Spanish.

Model	Train	Acc	Prec	Rec	F1
LogReg	Zh	88.3	76.8	90.9	83.3
LogReg*	En	87.2	76.9	85.6	81.0
BERT	Zh	91.3	81.2	94.5	87.3
BERT*	En	92.4	88.6	87.6	88.1
mBERT	En	88.2	**91.9**	69.0	78.8
mBERT	Zh	90.9	80.2	95.0	87.0
mBERT	En+Zh	**93.2**	86.2	93.6	**89.7**
mBERT	ALL	91.7	81.7	**95.5**	88.0

(b) Results on Chinese.

Model	Train	Acc	Prec	Rec	F1
LogReg	Fr	89.4	77.8	93.6	84.9
BERT	Fr	**95.0**	89.6	**95.4**	**92.4**
mBERT	En	88.9	**91.4**	72.1	80.6
mBERT	Fr	92.1	82.6	**95.4**	88.6
mBERT	En+Fr	93.6	86.4	94.9	90.5
mBERT	ALL	94.0	89.8	91.6	90.7

(c) Results on French.

Model	Train	Acc	Prec	Rec	F1
LogReg	De	85.2	69.5	95.1	80.4
BERT	De	92.4	83.3	**95.5**	88.9
mBERT	En	81.2	**97.1**	42.4	59.0
mBERT	De	91.1	80.4	95.4	87.3
mBERT	En+De	91.9	82.9	93.9	88.0
mBERT	ALL	**93.0**	86.0	93.4	**89.5**

(d) Results on German.

Model	Train	Acc	Prec	Rec	F1
LogReg	Ja	86.5	71.9	94.7	81.7
BERT	Ja	91.2	82.3	92.3	87.0
mBERT	En	83.0	**93.2**	50.5	65.5
mBERT	Ja	92.4	83.6	94.7	88.8
mBERT	En+Ja	**92.8**	84.2	95.3	**89.4**
mBERT	ALL	90.7	79.4	**95.7**	86.8

(e) Results on Japanese.

Model	Train	Acc	Prec	Recall	F1
LogReg	It	88.5	76.8	91.7	83.6
BERT	It	**93.7**	85.5	96.5	**90.7**
mBERT	En	83.5	**91.1**	53.6	67.5
mBERT	It	92.8	84.6	94.8	89.4
mBERT	En+It	91.7	81.1	**96.6**	88.2
mBERT	ALL	92.7	84.3	94.6	89.2

(f) Results on Italian.

Table 6: Results on different target languages. LogReg and BERT are trained on the translated target-language documents. LogReg* and BERT* are trained on English and applied on test reviews by translating the corresponding text from the target language to English. mBERT is trained with various configurations.

Detailed non-English results. Table 6 shows detailed results on non-English datasets. For Spanish and Chinese we evaluated an additional baseline where test reviews are translated to English and considered by LogReg ("Logreg*" baseline) or BERT ("BERT*" baseline) that were trained on English reviews only. This approach is less effective, as well as more expensive than the other approaches: to deploy in health departments, it would require each new test review to be translated to English. While BERT has the highest F1 score on average over all approaches, mBERT has higher recall than BERT on most non-English target languages.

We detect reviews mentioning foodborne illness. To demonstrate the potential of our approach for detecting foodborne illness, we ran mBERT on unlabeled restaurant reviews from the NYC Area, LA Area, and the Yelp Challenge dataset. Table 7 shows two examples that were classified as "Sick"

by our classifier. Translating those two reviews to English and applying LogReg (trained in English) led to a (wrong) "Not Sick" prediction, possibly because the translated reviews are not matching the training distribution for LogReg.

5 Discussion and Future Work

We presented our cross-lingual learning method for scaling foodborne illness detection to languages beyond English without extra annotations for non-English languages. As most reviews do not discuss foodborne illness, it is challenging to create proper evaluation datasets for all languages.

In our preliminary experiments, we evaluated our approach on non-English languages by translating labeled test reviews from English to other languages. A caveat of this evaluation approach is that complaints of foodborne illness in native-language reviews may be expressed differently than

Spanish	**Original (Es) text:** Definitivamente mi peor experiencia, me intoxique con un ostra mala, llevo 4 días en muy malas condiciones, por favor tengan cuidado, los ostiones y mariscos no se pueden comer en cualquier lugar, yo aprendi por las malas, espero que mi experiencia le sirva a alguien **mBERT (train: ALL) prediction:** "Sick" ✓
	Translated (En) text: Definitely my worst experience, I got intoxicated with a bad oyster, I have been in very bad conditions for 4 days, please be careful, the oysters and shellfish cannot be eaten anywhere, I learned through the bad ones, I hope my experience will serve you someone **LogReg (train: En) prediction:** "Not Sick" ✗
Chinese	**Original (Zh) text:** 装修和服务都还不错，但味道极差：底料没有味道，我们自己加了几次盐和料才勉强能吃，菜品也非常不新鲜。一顿饭吃得我们四个人都很生气，然后回家三个人都拉肚子。绝对不会再去吃。Avoid!! **mBERT (train: ALL) prediction:** "Sick" ✓
	Translated (En) text: The decoration and service are good, but the taste is very bad: the base material has no taste, we added salt several times to make it barely edible, and the dishes are very fresh. Four of us were angry at a meal, and then all three got diarrhea. Will never eat again. Avoid !! **LogReg (train: En) prediction:** "Not Sick" ✗
German	**Original (De) text:** Wir haben hier 2 bowls mit Steak und einen Burger gegessen. Für unverschämte 70,03$ gab es recht kleine und nicht wirklich gute Portionen (besonders die bowls). Nachdem mein Sohn von der Bowl gegessen hat, musste er brechen. Auch meiner Tochter und mir war schlecht. Der Service wirkte lieblos und desinteressiert. Die bowls kamen gerade mal lauwarm an unseren Tisch und die Chips vom Burger schmeckten nach nichts. Nicht zu empfehlen!!! **mBERT (train: ALL) prediction:** "Sick" ✓
	Translated (En) text: We ate 2 bowls of steak and a burger here. For outrageous $70.03 there were quite small and not really good portions (especially the bowls). After my son ate from the bowl, he had to break. My daughter and I were also bad. The service seemed careless and uninterested. The bowls just came to our table lukewarm and the chips from the burger didn't taste like anything. Not recommendable!!! **LogReg (train: En) prediction:** "Not Sick" ✗

Table 7: Examples of Spanish, Chinese and German restaurant reviews in our dataset classified as "Sick" and their translations to English.

in automatically translated reviews and thus, performance numbers may not be fully indicative of performance in native reviews. Therefore, an important next step is to create better evaluation datasets.

Our exploratory results show that training mBERT in multiple languages jointly is more effective than training mBERT on English (zero-shot approach) or the target-language only. On average across languages mBERT is outperformed by monolingual BERT trained on (translated) target-language documents. On the other hand, deploying mBERT in health departments for daily inspections would be easier as it would not require extra pre-processing steps such as language detection

that may introduce errors. Also, we showed that mBERT could potentially be applied for languages that were not seen in the training set, without extra translation efforts.

As another interesting direction for future work, we plan to evaluate the cross-lingual transfer approach of Karamanolakis et al. (2020), which applies even for low-resource languages that are not supported by mBERT or for which machine translation systems are not available. We also plan to extend our system for predicting which languages to use as source languages to achieve good performance on a target language (Lin et al., 2019).

Acknowledgments

We thank the anonymous reviewers for their constructive feedback. This material is based upon work supported by the National Science Foundation under Grant No. IIS-15-63785.

References

Jacob Devlin, Ming-Wei Chang, Kenton Lee, and Kristina Toutanova. 2019. BERT: Pre-training of deep bidirectional transformers for language understanding. In *Proceedings of the 2019 Conference of the North American Chapter of the Association for Computational Linguistics: Human Language Technologies*.

Thomas Effland, Anna Lawson, Sharon Balter, Katelynn Devinney, Vasudha Reddy, HaeNa Waechter, Luis Gravano, and Daniel Hsu. 2018. Discovering foodborne illness in online restaurant reviews. *Journal of the American Medical Informatics Association*.

Stephan Gouws and Anders Søgaard. 2015. Simple task-specific bilingual word embeddings. In *Proceedings of the 2015 Conference of the North American Chapter of the Association for Computational Linguistics: Human Language Technologies*.

Jenine K Harris, Leslie Hinyard, Kate Beatty, Jared B Hawkins, Elaine O Nsoesie, Raed Mansour, and John S Brownstein. 2018. Evaluating the implementation of a Twitter-based foodborne illness reporting tool in the City of St. Louis Department of Health. *International Journal of Environmental Research and Public Health*, 15(5).

Jenine K Harris, Raed Mansour, Bechara Choucair, Joe Olson, Cory Nissen, and Jay Bhatt. 2014. Health department use of social media to identify foodborne illness-Chicago, Illinois, 2013-2014. *Morbidity and Mortality Weekly Report*, 63(32):681–685.

Giannis Karamanolakis, Daniel Hsu, and Luis Gravano. 2019. Weakly supervised attention networks for fine-grained opinion mining and public health. In *Proceedings of the 5th Workshop on Noisy User-Generated Text*.

Giannis Karamanolakis, Daniel Hsu, and Luis Gravano. 2020. Cross-lingual text classification with minimal resources by transferring a sparse teacher. In *Proceedings of the 2020 Findings of Empirical Methods in Natural Language Processing*.

Kaliyaperumal Karthikeyan, Zihan Wang, Stephen Mayhew, and Dan Roth. 2019. Cross-lingual ability of multilingual BERT: An empirical study. In *International Conference on Learning Representations*.

Yu-Hsiang Lin, Chian-Yu Chen, Jean Lee, Zirui Li, Yuyan Zhang, Mengzhou Xia, Shruti Rijhwani, Junxian He, Zhisong Zhang, Xuezhe Ma, et al. 2019. Choosing transfer languages for cross-lingual learning. In *Proceedings of the 57th Annual Meeting of the Association for Computational Linguistics*.

Telmo Pires, Eva Schlinger, and Dan Garrette. 2019. How multilingual is multilingual BERT? In *Proceedings of the 57th Annual Meeting of the Association for Computational Linguistics*.

Anna Rogers, Olga Kovaleva, and Anna Rumshisky. 2020. A primer in BERTology: What we know about how BERT works. *arXiv preprint arXiv:2002.12327*.

Sebastian Ruder, Ivan Vulić, and Anders Søgaard. 2019. A survey of cross-lingual word embedding models. *Journal of Artificial Intelligence Research*, 65:569–631.

Adam Sadilek, Henry A Kautz, Lauren DiPrete, Brian Labus, Eric Portman, Jack Teitel, and Vincent Silenzio. 2016. Deploying nEmesis: Preventing foodborne illness by data mining social media. In *Proceedings of the AAAI Conference on Artificial Intelligence*.

Shijie Wu and Mark Dredze. 2019. Beto, Bentz, Becas: The surprising cross-lingual effectiveness of BERT. In *Proceedings of the 2019 Conference on Empirical Methods in Natural Language Processing and the 9th International Joint Conference on Natural Language Processing*.

Detection of Mental Health Conditions from Reddit via Deep Contextualized Representations

Zhengping Jiang
Computer Science Dept.
Columbia University
zj2265@columbia.edu

Sarah Ita Levitan
Computer Science Dept.
Hunter College, CUNY
sarah.levitan@hunter.cuny.edu

Jonathan Zomick
Psychology Dept.
Hofstra University
jzomick1@pride.hofstra.edu

Julia Hirschberg
Computer Science Dept.
Columbia University
julia@cs.columbia.edu

Abstract

We address the problem of automatic detection of psychiatric disorders from the linguistic content of social media posts. We build a large scale dataset of Reddit posts from users with eight disorders and a control user group. We extract and analyze linguistic characteristics of posts and identify differences between diagnostic groups. We build strong classification models based on deep contextualized word representations and show that they outperform previously applied statistical models with simple linguistic features by large margins. We compare user-level and post-level classification performance, as well as an ensembled multiclass model.

1 Introduction

Global prevalence of mental disorders has been estimated at 29.2% in a meta-study of 174 surveys across 63 countries (Steel et al., 2014). Mental illness is one of the leading causes of disability globally and the costs of mental health treatment have run into the trillions of dollars (Organization et al., 2014; Vigo et al., 2016; Patel et al., 2018). Additionally, individuals suffering from mental illness are estimated at forming 14.3% of deaths worldwide, significantly higher than a control population (Walker et al., 2015). Limited mental health resources and funding have necessitated new approaches to addressing the global impact of this problem. However, early detection of mental illness and early intervention have shown promising results for improving treatment and long-term outcome results for many psychiatric disorders; these have the potential to reduce the costly burden that mental illness has placed on our society as well as our global economies (Bird et al., 2010; Treasure and Russell, 2011; De Girolamo et al., 2012; Murru and Carpiniello, 2018).

Advances in artificial intelligence in general and computational linguistics in particular have made important contributions to detecting and predicting mental illness among the population, particularly in social media (Guntuku et al., 2017; Wongkoblap et al., 2017). Using computational linguistics, researchers have been able to leverage the widespread use of social media to analyze large, publicly available datasets for identifying linguistic markers of mental illness. To date, unique linguistic markers and patterns have been identified for several psychiatric conditions, such as major depressive disorder (MDD)(De Choudhury et al., 2013; Vedula and Parthasarathy, 2017), general anxiety disorder (GAD) (Shen and Rudzicz, 2017), bipolar disorder (BD) (Huang et al., 2017; Sekulić et al., 2018), eating disorders (ED) (Mohammadi et al., 2019; Naderi et al., 2019), schizophrenia (SZ) (Mitchell et al., 2015; Birnbaum et al., 2017; Zomick et al., 2019), obsessive compulsive disorder (OCD) (Coppersmith et al., 2015a), post-traumatic stress disorder (PTSD) (Coppersmith et al., 2014), as well as others (Coppersmith et al., 2015a). Linguistic findings have spanned various domains of language, including the use of pronouns, emotion words, tentative language, tangentiality, punctuation, and content analysis. The majority of these models have been developed to successfully predict if a given user has self-disclosed receiving a diagnosis for a psychiatric condition and is currently suffering with mental illness.

However, much of this previous research on social media and mental health has focused on comparing users with particular disorders with control users. In this work we expand this focus to compare across a wide set of common disorders. This is directly applicable to real-world diagnostic scenarios, where clinicians select a diagnosis from a large set of disorders, rather than simply diagnosing an individual as healthy or not. In addition, prior work

Proceedings of the 11th International Workshop on Health Text Mining and Information Analysis, pages 147–156
November 20, 2020. ©2020 Association for Computational Linguistics
https://doi.org/10.18653/v1/P17

has focused on data collection and analysis, with less emphasis on building strong predictive models. In this work, we apply state-of-the-art neural network models developed for other natural language tasks to the problem of mental health detection from social media.

2   Related Work

In recent years, there has been increased interest in the NLP community in the automatic detection of psychiatric conditions from language. Many researchers have focused on analyzing vast amounts of language from social media posts to study mental health (Birnbaum et al., 2017; Coppersmith et al., 2015a; Mitchell et al., 2015). With the advent of social media, many people who suffer from various forms of mental illness have found a sense of community and support, and these platforms offer a mode of expression for discussing their experiences openly online. Additionally, many online platforms allow users to post anonymously, giving them a sense of security and anonymity to discuss their experiences and struggles without the fear of being stigmatized or discriminated against (Balani and De Choudhury, 2015; Berry et al., 2017; Highton-Williamson et al., 2015).

In order to analyze language patterns related to various disorders from social media data, researchers have developed innovative approaches for automatically labeling this data. (Coppersmith et al., 2014) developed a widely used approach for gathering data for a range of psychological disorders, using regular expressions to identify public self-disclosures of diagnoses on social media. They tested this approach using Twitter data and collected a dataset of tweets from individuals with bipolar, depression, PTSD, SAD, and a control group. They analyzed several linguistic features across conditions using a clustering algorithm and built predictive classifiers to distinguish between diagnosed and control users. (Cohan et al., 2018a) expanded this approach to study a larger set of disorders using Reddit data. Reddit is one of the fastest growing and widely used social media platforms, averaging over 330 million active monthly users, and, as of 2018, was the fourth most visited website in the US (Hutchinson, 2018). Unlike Twitter, Reddit imposes no limits on the length of posts, enabling an analysis of longer language samples. In addition, Reddit is composed of subreddits, which are forums dedicated to specific topics, and

there are many subreddits related to specific mental health conditions. They collected a large dataset of Reddit posts and analyzed linguistic features between different conditions and a control group. They also trained binary classifiers to distinguish between each condition and the control.

Our work directly builds on this prior work. Following (Coppersmith et al., 2014) and (Cohan et al., 2018a), we collect a large expanded dataset of Reddit posts. Unlike prior work, we do not focus on pairwise analyses of linguistic features between conditions and the control group; rather, we compare features between conditions to highlight important differences that can distinguish between various disorders. While others have trained simple predictive models of these disorders, we instead use state-of-the-art deep contextualized models that have been highly successful across several NLP tasks. Our work makes important contributions to the problem of mental health detection from social media data and provides insights for others to further build on this work.

3   Data Collection

We focus in this study on 8 mental health conditions: schizophrenia (SZ), borderline personality disorder (BPD), post-traumatic stress disorder (PTSD), eating disorder (ED), major depression disorder (MDD), general anxiety disorder (GAD) and bipolar disorder. While datasets for many of these conditions have been collected on varying scales, to the best of our knowledge our dataset includes the largest cohort of users whose posts have been collected for many of these conditions. To build this cohort, we collect users with self-identified mental health conditions from Reddit using the Pushshift API[1]. We search for users in mental health related subreddits and use keywords to search for mental health related words. Our distant labeling approach is further explained below, in Section 3.1. We also identify a group of control users who do not have any of the targeted conditions. We first collect a large scale user pool by scraping posts from common subreddits like r/AskReddit, and filter the control users by process described in subsection 3.2. The number of posts collected in each condition is shown in Table 1 and the number of users whose posts were collected in each is shown in Table 2.

---

[1] https://github.com/pushshift/api

	Post Number	Avg. Token
SZ	1084k	43.7
BPD	1629k	43.6
PTSD	2169k	46.1
ED	396k	43.0
MDD	1585k	42.9
GAD	3047k	42.2
OCD	1813k	38.6
Bipolar	5819k	40.5
Total	17.5m	42.0

Table 1: Dataset Statistics (Posts)

	User Num	Unique	Clf.
SZ	2134	1741	1175
BPD	4695	3430	2275
PTSD	5294	3840	2666
ED	1005	752	514
MDD	3183	1832	1360
GAD	4958	3155	2388
OCD	4151	3140	2211
Bipolar	11186	9524	6420
Total	35606	27214	19009

Table 2: Dataset Statistics (Users), where **Unique** column figures correspond to number of users without comorbidity issue and **Clf.** column figures correspond to number of users we use for our classification task.

3.1 Distant Labeling

We generally follow the self-identification technique previously employed in (Mitchell et al., 2015; Coppersmith et al., 2015a; Cohan et al., 2018a). Specifically, we construct separate regular expressions for self-identification checking and condition resolution. We use 2-way human annotation to verify the performance of our labeling algorithm. Our second version of labeling algorithm achieves high precision (over .95) when tested on a held-out validation set. We found that posts directly identifying with "eating disorder" are scarce, so we collapse identification with "anorexia", "arfid", "bulimia" and "binge" into a single category for "eating disorder". We also calculate comorbidity statistics for our extracted user set as is shown in Figure 1, and have found it correlated well with statistics previously reported (Coppersmith et al., 2015a; Cohan et al., 2018b).

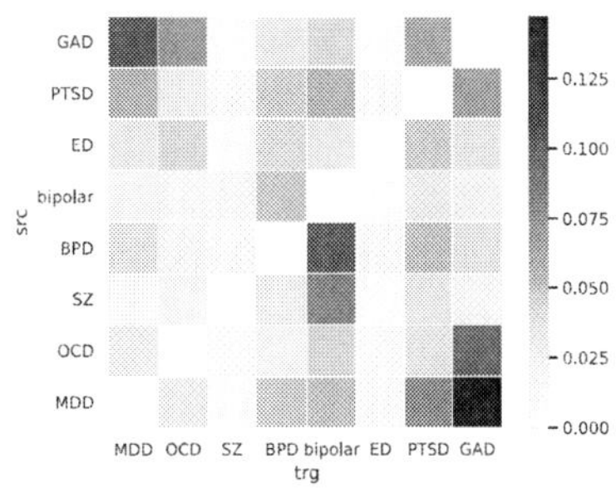

Figure 1: Comorbidity matrix of our dataset, each ceil corresponds to the portion of users of the src condition that have the trg condition.

3.2 Preprocessing

Following Cohan et al.(2018b) we do not include in our classification any control user that has any *sensitive post*, defined as either (1) containing mental health related keywords or (2) posted in a mental health related subreddit. In addition, under the (**CL**) condition of our classification experiments (described below in Section 6), we remove these sensitive posts from mental group users. For post level preprocessing, we replace emojis with descriptive text using the demoji package[2], normalize html characters like "", "&" and " " etc., and mask out url, email and subreddit references with regular expression.

4 Linguistic Indicators of Mental Health

After collecting and preprocessing the data, we analyzed linguistic characteristics of mental health using Linguistic Inquiry and Word Count (Pennebaker et al., 2015). LIWC is a text analysis program that computes word counts for semantic classes and structural features. It relies on an internal dictionary that maps words to psychologically motivated categories. These include standard linguistic features (e.g. percentage of words that are pronouns, articles), markers of psychological processes (e.g. affect, social, cognitive words), and punctuation categories (e.g. periods, commas). LIWC dimensions have been used in many studies to predict outcomes including personality (Pennebaker and King, 1999), deception (Newman et al., 2003), and health (Pennebaker et al., 1997). We extracted 73 features using LIWC 2015; a full description of these features is found

---

[2]https://pypi.org/project/demoji/

in (Pennebaker et al., 2015). To construct a single feature vector per user, we concatenated all posts per user and then extracted the LIWC features from the combined posts and performed length normalization.

Prior work on identifying linguistic indicators of mental health has compared LIWC features from users' individual disorders with healthy control users. However, it is often unclear whether the findings are specific to the disorder, or if they are indicative of mental disorders more generally. For example, in pairwise analyses, personal pronoun usage has been found to be increased in individuals with schizophrenia (Zomick et al., 2019). However, this pattern might or might not be specific to schizophrenia, but may be indicative of other mental disorders as well. Because of this gap in prior work, we began by comparing LIWC features directly across the 8 diagnostic groups and the control group.

Figure 2 shows a heatmap of the z-score normalized average LIWC features across users in each group. The x-axis shows the 8 diagnostic groups and the control group, and the y-axis shows the normalized LIWC feature values. The color of each cell indicates whether the scaled value is high (blue), low (red) or average (white). As shown in this figure, the control group has the greatest number of red features, or LIWC features which have a low frequency. It is clear from the figure that the 8 diagnostic groups have different language usage patterns from the control group, and particularly show a higher frequency for several linguistic dimensions. Further, there seem to be several interesting similarities and differences in linguistic patterns across the diagnostic groups. To further investigate these differences, we ran one-way ANOVAs comparing each LIWC feature across the 8 diagnostic groups and the control group. To correct for family-wise type I errors, we used Bonferroni correction. The results indicated that there were significant differences across groups for all 73 LIWC variables. We ran Tukey posthoc tests to identify which pairs of conditions were most similar and most different. Because of limited space, we focus here on the linguistic dimensions with the greatest variance among the groups, indicated by the highest F-statistics. These categories were *anx*, the use of anxiety words ($F(8, 24442) = 531.911$, $p<.0001$), and *I*, the use of the first person singular pronoun ($F(8, 24442) = 438.738$, $p<.0001$). Figure

3 shows the results of the posthoc analysis. Pairwise comparisons among psychiatric conditions revealed several interesting findings. While each condition differed significantly for both features when compared with the control group (users in the control group were significantly less likely to use anxiety related words and "I" when compared to each condition), when comparing between the psychiatric conditions differences varied. For example, users with SZ were significantly less likely to use anxiety related words in comparison with other groups. Another interesting finding was that users with BPD used 1st person singular pronouns significantly more than all other psychiatric conditions with the exception of ED. These findings shed light on linguistic variation across different psychiatric conditions, and provide further motivation for developing methods to distinguish between individuals with these disorders by leveraging social media posts.

5   Methods for Classification Experiments

Having identified significant differences in linguistic features between the disorders, and between the control users, we next explore several classification methods for automatically identifying different mental health conditions. Previous efforts to identify such conditions in Reddit have primarily employed simple logistic regression or SVMs using a bag-of-words representation or LIWC features, and some have explored RNN/CNN based text encoder models (Coppersmith et al., 2014, 2015a,b; Cohan et al., 2018b; Sekulic and Strube, 2019). However, recent advances in contextual representations like ELMo (Peters et al., 2018) and BERT (Devlin et al., 2018), which have enabled substantial performance increases across many NLP task, have not been well integrated into mental health identification tasks. This is due to model size and scalability issues of the large number of posts generated by a user. In this work we focus on methods utilizing contextual representations for mental health identification and compare their effectiveness to a logistic regression baseline trained on LIWC features. We present an attention-based model using BERT representations as input features, as well as a REALM-like model (Guu et al., 2020) inspired by recent advances in open domain question-answering. All of these models are trained for a user-level classification task, to detect whether a user has a particular

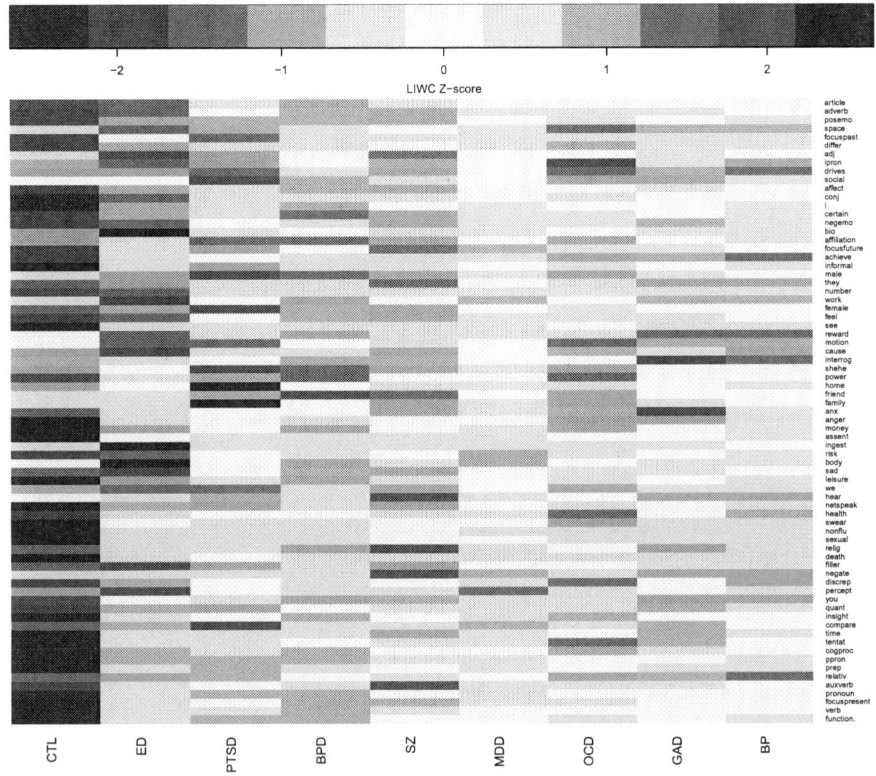

Figure 2: Scaled LIWC features across 8 diagnostic groups and control group.

diagnosis, based on an aggregated representation of their posts. In addition, we conduct post-level classification experiments with classic BERT fine-tuning settings. This experiment is done to discover the importance of global context in classification. Finally, in addition to these binary classifiers (diagnosis vs. control), we ensemble all our binary classification models as a multi-label classifier among different diagnostic groups, which is the ultimate goal for the application of this work.

For user-level classification, we select users not belonging to the co-morbidity group as a control group. To reduce the size of the data, we remove posts less than 50 characters long from both the mental group and the control group, as we hypothesize that these may not provide enough information for classification. Also we do not include control users with fewer than 20 non-sensitive posts. When pairing with mental health users, We select control users who have a similar total number of posts, who do not have mental health sensitive posts, and who have at least some subreddit overlap with the mental health users, as described by Cohan et al. (2018b). We consider two experimental settings: for the **CL** (clean) experiments we exclude all mental health sensitive posts for mental health users, and for the **UNCL** (unclean) experiments we include these posts for their corresponding users. The intuition is that under the **UNCL** setting, our model should be able to make predictions based on some explicit semantic triggers, thus resulting in better performance. However, under the **CL** setting, the model may rely on underlying syntactic differences that may generalize better than explicit semantic features.

Below we describe the Attention-Based model and the REALM model that we adapt for this work.

5.1 Attention-Based Model

A direct solution to the scalability issue with this data is to restrain the gradient update in a model to a small portion of the model parameters. We propose to use pre-trained BERT model from Hugging Face (Wolf et al., 2019) to encode every post;

Mean Difference in Overall ANX Usage Between Groups (i − j)

i \ j	Ctrl	Bipolar	BPD	ED	GAD	MDD	OCD	PTSD	SZ
Ctrl		-7.9**	-9.7**	-1.1**	-2.0**	-6.6**	-1.6**	-10.0**	-5.4**
Bipolar	.008**		-1.8**	-3.5**	-1.2**		-7.7**	-2.0**	2.5**
BPD	.020**	.002*			-9.9**	3.2**	-5.8**		4.3**
ED	.020**				-8.2**	4.9**	-4.2**		6.0**
GAD	.016**	-.002**	-.004**	-.005**		1.3**	4.0**	9.7**	1.4**
MDD	.014**	-.005**	-.007**	-.007**	-.002*		-9.0**	-3.4**	
OCD	.017**		-.003**	-.004**		.003**		5.6**	1.0**
PTSD	.018**		-.002**		.002*	.004**			4.5**
SZ	.015**	-.003**	-.005*	-.006**				-.003**	

*p < .01, **p < .001

Figure 3: LIWC analysis across 8 diagnostic groups and control group for anxiety (ANX, in blue) and singular personal pronoun usage (I, in orange). Only significant results are displayed in this table.

we then averaged representation of all positions as a pooling result to build an attention-based classifier (Bahdanau et al., 2014; Sutskever et al., 2014) over all the post-level representations for a single user[3]. This resembles the settings of many "probing tasks" (Hewitt and Manning, 2019) used to investigate whether BERT embeddings encode useful linguistic information about a user's mental health condition.

5.2 REALM-like Models

Guu et al. (2020) propose Retrieval-Augmented Language Model pretraining to augment a pretrained LM as a textual knowledge retriever. To tackle the scalability issue of retrieving over large corpora, a retrieval encoder parameterized by θ tuned over their top-k retrieval results is used to encode all documents in the textual knowledge corpus. Guu et al. (2020) shows by gradient analysis that a document z will receive a positive update if the estimated probability of a correct answer y based on z is better than the expectation over all documents in the textual knowledge corpus. To adapt this REALM model to our task, we reformulate our classification problem as a "retrieve-then-predict" pipeline similar to Open Domain Question Answering (ODQA). Specifically, given a user's total set of posts $\mathcal{Z}$, we first select the top-k posts $\{z_1, \ldots, z_k\}$ that are most helpful in predicting the user's mental health condition and we base our prediction only on these posts. Unlike in ODQA we have a ques-

tion x that can be utilized for relevant document selection, so we now use a trainable attention head to calculate the retrieval probability $p(z)$. Thus the probability of a user having condition y can be factorized as:

$$p(y) = \sum_{z \in \mathcal{Z}} p(y|z)p(z) \tag{1}$$

where

$$p(z) = \frac{h^T \text{Embed}_{\text{doc}}(z)}{\sum_{z'} h^T \text{Embed}_{\text{doc}}(z')} \tag{2}$$

Here, $\text{Embed}_{\text{doc}}(\cdot)$ is implemented as a BERT-style transformer parameterized by θ and $p(y|z)$ is implemented as a BERT-based classifier parameterized by ϕ. When training, we first index all user posts with $\text{Embed}_{\text{doc}}(z)$ using our model θ, and jointly tune θ and ϕ and h on the top-k user posts w.r.t. $p(z)$. For every several epochs, we re-index all posts with tuned parameter θ'.

5.3 Experimental Settings

For REALM-like models we update the user corpus index every 5 epochs. At every step we use the top 10 documents to tune the model for each user, and we set the learning rate for the attention-based classifier to 1e-3, and the learning rate for BERT parameters to 1e-5. For the attention-based model, we set the learning rate for the classifier to 1e-3 and keep the BERT parameters frozen. For our post-level classification model we set the learning rate in the same way as in the REALM-like model. Note that when fine-tuning the BERT-based model, we pool the sentence representation with $[CLS]$ token, unlike the non-tunable model (BERT-ATT)

---

[3]Note that we are not using the $[CLS]$ (start sequence token) as the pooling result. This is because in the pretraining model it is used for next sentence prediction and thus is not an ideal semantic representation. The specific structure of BERT-ATT model does not allow the representation to be tuned.

where we average across all positions to get the pooling result. In all cases except for our LIWC-feature-based logistic regression model, we use a held-out development set for model selection; for LIWC-based regression we run a parameter grid search using cross-validation on the training set. For the multi-label classification experiments, we ensemble the best model set under the **CL** setting as the multi-label classifier.

6 Mental Health Detection Results

In this section we present the results for user-level and post-level binary classification, and for the multiclass ensemble classification.

6.1 User-Level Classification

Table 3 shows the user-level binary classification results, comparing the BERT-attention model, REALM model, and the LIWC logistic regression model, under both **CL** and **UNCL** settings. We find that, under the **UNCL** setting, the REALM-like model consistently achieves the highest accuracy for all diagnostic groups. Under the **CL** setting, tuning a classifier over the original BERT representation achieves better results for all groups. This is probably because the REALM model makes its predictions using only the top-10 segments retrieved from all of a user's posts, while the BERT-ATT model is able to attend to all the posts at once. This result aligns well with the intuition that linguistic traits for mental health conditions should be global, and may be more difficult to determine from a small portion of posts – especially when posts containing sensitive keywords are removed. The BERT-ATT model performs best for the bipolar category (CL-F1: .879; UNCL-F1: .931) for which we have the largest user group, indicating the importance of obtaining large scale datasets for the success of deep mental health detection. In all cases, our results suggest that contextualized representation is a better feature for mental health prediction compared with LIWC features, but is also more likely to model shallow semantic traits.

6.2 Post-Level Classification

To create a balanced set comparable to user level classification, we sample 50,000 mental group posts and 50,000 control group posts as the training set, and 5,000 + 5,000 posts for dev and test. Table 4 shows the post-level binary classification results. This post-level classification result ranges

from an F1 of .596 for MDD to an F1 of .736 for ED, substantially lower than the user-level classification performance. This suggests that linguistic signals related to mental health problems do not appear in all posts of a mental group user. However, model performance exhibits similar trends when we compare post-level with user-level classifications LIWC features, indicating that the ED subset is the easiest and MDD the hardest: this result may mean that linguistic traits for ED have a broader coverage among user posts while for MDD the scope is probably smaller. This is consistent with the results reported by Coppersmith et al. (2015a).

6.3 Multi-label Model Ensemble

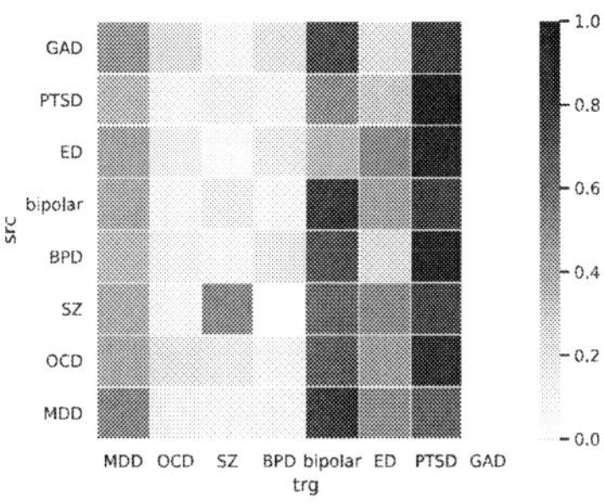

Figure 4: Multilabel ensemble experiments result. The overall result is $F-1_{micro} = 0.2175$ and $F-1_{macro} = 0.195$. Each cell representing the portion of users with gold label in src and predicted label in trg by our ensemble classifier.

As the BERT-ATT model performs best under **CL** setting, we ensemble all BERT-ATT model as the multi-label classifier. Again, to create a balanced testing set, we sample 100 users from each condition group's test set. We then predict the user condition by selecting the label with the highest score from the model. With this naive ensemble method we achieve $F - 1_{micro} = 0.2175$ and $F - 1_{macro} = 0.195$. The fact that these results are only slightly above the random baseline (.125) indicates that, though under binary settings deep contextualized word representation is a strong feature, the model is not well calibrated, (DeGroot and Fienberg, 1983; Niculescu-Mizil and Caruana, 2005) as is often the case for modern deep networks (Guo et al., 2017). To see whether there are any identifiable patterns in the errors, we plot the prediction heatmap for the multi-label classification task, as shown in Figure 4. We find that there is a discrepancy of confidence between dif-

		CL			UNCL		
		LIWC	**BERT-ATT**	**REALM**	**LIWC**	**BERT-ATT**	**REALM**
SZ	Acc.	0.668	**0.808**	0.727	0.757	0.885	**0.97**
	F-1	0.685	**0.812**	0.766	0.775	0.898	**0.973**
BPD	Acc.	0.729	**0.884**	0.822	0.792	0.89	**0.995**
	F-1	0.716	**0.875**	0.82	0.782	0.877	**0.995**
PTSD	Acc.	0.703	**0.872**	0.712	0.764	0.892	**0.979**
	F-1	0.694	**0.877**	0.728	0.758	0.885	**0.978**
ED	Acc.	0.75	**0.873**	0.825	0.843	0.877	**0.99**
	F-1	0.732	**0.882**	0.838	0.85	0.887	**0.99**
MDD	Acc.	0.67	**0.833**	0.822	0.702	0.91	**0.965**
	F-1	0.663	**0.843**	0.819	0.719	0.908	**0.965**
GAD	Acc.	0.799	**0.834**	0.758	0.845	0.855	**0.988**
	F-1	0.789	**0.799**	0.764	0.835	0.847	**0.989**
OCD	Acc.	0.721	**0.865**	0.746	0.815	0.884	**0.973**
	F-1	0.71	**0.872**	0.707	0.807	0.875	**0.974**
Bipolar	Acc.	0.699	**0.872**	0.82	0.778	0.926	**0.983**
	F-1	0.692	**0.879**	0.813	0.773	0.931	**0.982**

Table 3: User-level BERT classification results. The best result for a mental group under specific settings is in bold.

	Accu.	**F-1**
SZ	0.628	0.614
BPD	0.689	0.689
PTSD	0.630	0.577
ED	0.708	0.736
MDD	0.567	0.596
GAD	0.675	0.683
OCD	0.627	0.65
Bipolar	0.598	0.615

Table 4: Post-level BERT Classification Results

ferent models, and this confidence neither strongly correlates with training data size nor with binary classification performance. Though we observe that cells on the main diagonal in general have a darker shade indicating a promising separation of feature sets that are useful in identifying their designated condition, the mislabeling distribution has little resemblance to the comorbidity distribution characterized in Figure 1. Further experimentation is needed to improve the multiclass ensemble classification performance.

7   Conclusions and Future Work

In this paper we collect and analyze a large scale dataset of social media posts from users various mental health conditions. We analyze linguistic characteristics of the posts, directly comparing the features of the various conditions. We build strong classification models based on deep contextualized representations and demonstrate that they outperform the LIWC feature based logistic regression baseline by a large margin. Although the LIWC feature representation is not as useful for classification, it is a useful representation for analysis of posts to gain insight about the differences between groups. Our experimental results show that linguistic traits for mental health detection are more easily recognized at the user-level and thus effectively aggregating post-level signals is crucial to accurate prediction. Also, we find that these contextualized representations rely heavily on semantic content and always perform better when semantic indicators are obvious. We also show that the prediction scores of our classification models, even the accurate ones, are not well calibrated and thus are not an accurate uncertainty estimator of mental health risk. These results call for a more interpretable model for mental health detection. Future research may look into the direction of learning better deep features and exploring additional classification paradigms to further improve performance for this impactful problem.

References

Dzmitry Bahdanau, Kyunghyun Cho, and Yoshua Bengio. 2014. Neural machine translation by jointly learning to align and translate. *arXiv preprint arXiv:1409.0473*.

Sairam Balani and Munmun De Choudhury. 2015. Detecting and characterizing mental health related self-disclosure in social media. In *Proceedings of the 33rd Annual ACM Conference Extended Abstracts on Human Factors in Computing Systems*, pages 1373–1378. ACM.

Natalie Berry, Fiona Lobban, Maksim Belousov, Richard Emsley, Goran Nenadic, and Sandra Bucci. 2017. # whywetweetmh: understanding why people use twitter to discuss mental health problems. *Journal of medical Internet research*, 19(4).

Victoria Bird, Preethi Premkumar, Tim Kendall, Craig Whittington, Jonathan Mitchell, and Elizabeth Kuipers. 2010. Early intervention services, cognitive–behavioural therapy and family intervention in early psychosis: systematic review. *The British Journal of Psychiatry*, 197(5):350–356.

Michael L Birnbaum, Sindhu Kiranmai Ernala, Asra F Rizvi, Munmun De Choudhury, and John M Kane. 2017. A collaborative approach to identifying social media markers of schizophrenia by employing machine learning and clinical appraisals. *Journal of medical Internet research*, 19(8):e289.

Arman Cohan, Bart Desmet, Andrew Yates, Luca Soldaini, Sean MacAvaney, and Nazli Goharian. 2018a. SMHD: a large-scale resource for exploring online language usage for multiple mental health conditions. In *Proceedings of the 27th International Conference on Computational Linguistics*, pages 1485–1497, Santa Fe, New Mexico, USA. Association for Computational Linguistics.

Arman Cohan, Bart Desmet, Andrew Yates, Luca Soldaini, Sean MacAvaney, and Nazli Goharian. 2018b. Smhd: a large-scale resource for exploring online language usage for multiple mental health conditions. *arXiv preprint arXiv:1806.05258*.

Glen Coppersmith, Mark Dredze, Craig Harman, and Kristy Hollingshead. 2015a. From adhd to sad: Analyzing the language of mental health on twitter through self-reported diagnoses. In *Proceedings of the 2nd Workshop on Computational Linguistics and Clinical Psychology: From Linguistic Signal to Clinical Reality*, pages 1–10.

Glen Coppersmith, Mark Dredze, Craig Harman, Kristy Hollingshead, and Margaret Mitchell. 2015b. Clpsych 2015 shared task: Depression and ptsd on twitter. In *Proceedings of the 2nd Workshop on Computational Linguistics and Clinical Psychology: From Linguistic Signal to Clinical Reality*, pages 31–39.

Glen Coppersmith, Craig Harman, and Mark Dredze. 2014. Measuring post traumatic stress disorder in twitter. In *Eighth international AAAI conference on weblogs and social media*.

Munmun De Choudhury, Michael Gamon, Scott Counts, and Eric Horvitz. 2013. Predicting depression via social media. In *Seventh international AAAI conference on weblogs and social media*.

Giovanni De Girolamo, J Dagani, R Purcell, A Cocchi, and PD McGorry. 2012. Age of onset of mental disorders and use of mental health services: needs, opportunities and obstacles. *Epidemiology and psychiatric sciences*, 21(1):47–57.

Morris H DeGroot and Stephen E Fienberg. 1983. The comparison and evaluation of forecasters. *Journal of the Royal Statistical Society: Series D (The Statistician)*, 32(1-2):12–22.

Jacob Devlin, Ming-Wei Chang, Kenton Lee, and Kristina Toutanova. 2018. Bert: Pre-training of deep bidirectional transformers for language understanding. *arXiv preprint arXiv:1810.04805*.

Sharath Chandra Guntuku, David B Yaden, Margaret L Kern, Lyle H Ungar, and Johannes C Eichstaedt. 2017. Detecting depression and mental illness on social media: an integrative review. *Current Opinion in Behavioral Sciences*, 18:43–49.

Chuan Guo, Geoff Pleiss, Yu Sun, and Kilian Q Weinberger. 2017. On calibration of modern neural networks. *arXiv preprint arXiv:1706.04599*.

Kelvin Guu, Kenton Lee, Zora Tung, Panupong Pasupat, and Ming-Wei Chang. 2020. Realm: Retrieval-augmented language model pre-training. *arXiv preprint arXiv:2002.08909*.

John Hewitt and Christopher D Manning. 2019. A structural probe for finding syntax in word representations. In *Proceedings of the 2019 Conference of the North American Chapter of the Association for Computational Linguistics: Human Language Technologies, Volume 1 (Long and Short Papers)*, pages 4129–4138.

Elizabeth Highton-Williamson, Stefan Priebe, and Domenico Giacco. 2015. Online social networking in people with psychosis: a systematic review. *International Journal of Social Psychiatry*, 61(1):92–101.

Yen-Hao Huang, Lin-Hung Wei, and Yi-Shin Chen. 2017. Detection of the prodromal phase of bipolar disorder from psychological and phonological aspects in social media. *arXiv preprint arXiv:1712.09183*.

Andrew Hutchinson. 2018. Reddit now has as many users as twitter, and far higher engagement rates. https://www.socialmediatoday.com/news/reddit-now-has-as-many-users-as-twitter-and-far-higher-engagement-rates/521789/. Accessed: 2019-03-10.

Margaret Mitchell, Kristy Hollingshead, and Glen Coppersmith. 2015. Quantifying the language of schizophrenia in social media. In *Proceedings of the 2nd workshop on Computational linguistics and clinical psychology: From linguistic signal to clinical reality*, pages 11–20.

Elham Mohammadi, Hessam Amini, and Leila Kosseim. 2019. Quick and (maybe not so) easy detection of anorexia in social media posts. In *CLEF (Working Notes)*.

Andrea Murru and Bernardo Carpiniello. 2018. Duration of untreated illness as a key to early intervention in schizophrenia: a review. *Neuroscience letters*, 669:59–67.

Nona Naderi, Julien Gobeill, Douglas Teodoro, Emilie Pasche, and Patrick Ruch. 2019. A baseline approach for early detection of signs of anorexia and self-harm in reddit posts. In *Proceedings of the CLEF 2019 Workshop*.

Matthew L Newman, James W Pennebaker, Diane S Berry, and Jane M Richards. 2003. Lying words: Predicting deception from linguistic styles. *Personality and social psychology bulletin*, 29(5):665–675.

Alexandru Niculescu-Mizil and Rich Caruana. 2005. Predicting good probabilities with supervised learning. In *Proceedings of the 22nd international conference on Machine learning*, pages 625–632.

World Health Organization et al. 2014. *Global status report on noncommunicable diseases 2014.* WHO/NMH/NVI/15.1. World Health Organization.

Vikram Patel, Shekhar Saxena, Crick Lund, Graham Thornicroft, Florence Baingana, Paul Bolton, Dan Chisholm, Pamela Y Collins, Janice L Cooper, Julian Eaton, et al. 2018. The lancet commission on global mental health and sustainable development. *The Lancet*, 392(10157):1553–1598.

James W Pennebaker, Ryan L Boyd, Kayla Jordan, and Kate Blackburn. 2015. The development and psychometric properties of liwc2015. Technical report, University of Texas at Austin.

James W Pennebaker and Laura A King. 1999. Linguistic styles: language use as an individual difference. *Journal of personality and social psychology*, 77(6):1296.

James W Pennebaker, Tracy J Mayne, and Martha E Francis. 1997. Linguistic predictors of adaptive bereavement. *Journal of personality and social psychology*, 72(4):863.

Matthew E Peters, Mark Neumann, Mohit Iyyer, Matt Gardner, Christopher Clark, Kenton Lee, and Luke Zettlemoyer. 2018. Deep contextualized word representations. *arXiv preprint arXiv:1802.05365*.

Ivan Sekulić, Matej Gjurković, and Jan Šnajder. 2018. Not just depressed: Bipolar disorder prediction on reddit. *arXiv preprint arXiv:1811.04655*.

Ivan Sekulic and Michael Strube. 2019. Adapting deep learning methods for mental health prediction on social media. *Proceedings of the 5th Workshop on Noisy User-generated Text (W-NUT 2019)*.

Judy Hanwen Shen and Frank Rudzicz. 2017. Detecting anxiety through reddit. In *Proceedings of the Fourth Workshop on Computational Linguistics and Clinical Psychology—From Linguistic Signal to Clinical Reality*, pages 58–65.

Zachary Steel, Claire Marnane, Changiz Iranpour, Tien Chey, John W Jackson, Vikram Patel, and Derrick Silove. 2014. The global prevalence of common mental disorders: a systematic review and meta-analysis 1980–2013. *International journal of epidemiology*, 43(2):476–493.

Ilya Sutskever, Oriol Vinyals, and Quoc V Le. 2014. Sequence to sequence learning with neural networks. In *Advances in neural information processing systems*, pages 3104–3112.

Janet Treasure and Gerald Russell. 2011. The case for early intervention in anorexia nervosa: theoretical exploration of maintaining factors. *The British Journal of Psychiatry*, 199(1):5–7.

Nikhita Vedula and Srinivasan Parthasarathy. 2017. Emotional and linguistic cues of depression from social media. In *Proceedings of the 2017 International Conference on Digital Health*, pages 127–136.

Daniel Vigo, Graham Thornicroft, and Rifat Atun. 2016. Estimating the true global burden of mental illness. *The Lancet Psychiatry*, 3(2):171–178.

Elizabeth Reisinger Walker, Robin E McGee, and Benjamin G Druss. 2015. Mortality in mental disorders and global disease burden implications: a systematic review and meta-analysis. *JAMA psychiatry*, 72(4):334–341.

Thomas Wolf, Lysandre Debut, Victor Sanh, Julien Chaumond, Clement Delangue, Anthony Moi, Pierric Cistac, Tim Rault, Rémi Louf, Morgan Funtowicz, et al. 2019. Huggingface's transformers: State-of-the-art natural language processing. *ArXiv*, pages arXiv–1910.

Akkapon Wongkoblap, Miguel A Vadillo, and Vasa Curcin. 2017. Researching mental health disorders in the era of social media: systematic review. *Journal of medical Internet research*, 19(6):e228.

Jonathan Zomick, Sarah Ita Levitan, and Mark Serper. 2019. Linguistic analysis of schizophrenia in reddit posts. In *Proceedings of the Sixth Workshop on Computational Linguistics and Clinical Psychology*, pages 74–83.